SAFER LIFTING
FOR
PATIENT CARE

*To all who care for the sick and disabled
in the hope that the use of this book will
lighten the physical load.*

SAFER LIFTING FOR PATIENT CARE

Margaret Hollis
MBE MCSP DipTP
Formerly Principal
Bradford Hospitals
School of Physiotherapy

with a Foreword by
P. R. Davis
Professor of Human
Biology and Health
Surrey University

Blackwell Scientific Publications
OXFORD LONDON EDINBURGH
BOSTON MELBOURNE

© 1981 by Blackwell Scientific
Publications
Editorial Offices:
Osney Mead, Oxford OX2 0EL
8 John Street, London WC1N 2ES
9 Forrest Road, Edinburgh EH1 2QH
52 Beacon Street, Boston,
    Massachusetts 02108 USA
99 Barry Street, Carlton,
    Victoria 3053 Australia

First published 1981

Printed and bound in Great Britain
by Henry Ling Ltd, Dorchester

DISTRIBUTORS

USA
    Blackwell Mosby Book
        Distributors
    11830 Westline
    Industrial Drive
    St Louis, Missouri 63141

Canada
    Blackwell Mosby Book
        Distributors
    120 Melford Drive,
    Scarborough
    Ontario M1B 2X4

Australia
    Blackwell Scientific Book
        Distributors
    214 Berkeley Street,
    Carlton
    Victoria 3053

British Library
Cataloguing in Publication Data

Hollis, Margaret
    Safer lifting for patient care.
    1. Transport of sick and wounded.
    I. Title
    610.73   RT87.T72

ISBN 0-632-008253

Contents

Tabulated List
of Lifts

Tabulated List of Lifts

Foreword

Whatever else those caring for the sick may need to do, they will be called upon frequently to lift, move or carry patients. Human beings are heavy, and fragile. When handling patients, the imposed load often approaches or exceeds the maximum levels advocated for many industrial workers. It is small wonder, therefore, that musculoskeletal disorders resulting from physical overstress are so common in members of these honoured professions. Recent studies have shown that back disorders may account for the loss to the nursing profession alone of some 20 per cent of entrants during their training period, and that the majority of registered nurses suffer from repeated attacks of incapacity from this cause, the frequency of attacks being highest in those specialties in which the nurses have to handle patients most frequently.

It is typical of the selflessness of the caring professions that, until very recently, much attention was given during training to the safety and comfort of the patient, but that little time was spared for the safety of the professional. Even now, in many training schedules, only one or two sessions are devoted to general patient handling.

There is now good evidence that, with adequate and proper training and the rigorous application of proper techniques, sick people can be handled in most circumstances with safety both for the patient and the operator. Equally, there is good evidence that this can only be achieved if training time is adequate, and the need for the rigorous application of proper methods is understood by all concerned. Patient handling devices can be obtained for use in many of those situations where simple handling is unsafe, but again, little or no training in their use may be available, and the design of the ward or sick room may make their use impracticable. For these reasons, I welcome this important book.

Margaret Hollis has wide experience of the needs of both the patients and of those whose lives are spent in helping them, and has made important contributions to the subject of handling safety. She has now used that experience and knowledge to produce a compendium of unaided patient handling techniques, with a review of the background sciences required to give the basis of a proper understanding of the subject. While a number of training pamphlets have been obtainable, one of the possible reasons for the inadequacy of training may have been the lack of a suitable textbook. It is to be hoped that Miss Hollis's new work will be used, as it should be, to fill that gap. Her gift for translating scientific incomprehensibility into plain English has been used to excellent effect, and her presentation of this important subject should commend itself to all who are faced with the need to handle patients.

Surrey University, 1981　　　　　*P. R. Davis*

Preface

During the 1950s I learned the Shoulder (Australian) Lift and incorporated it into my teaching to both physiotherapy and district nurse students. The latter always produced lifting problems borne of the very different conditions involved in nursing in the patient's own home. Indeed, problem-solving produced all the new lifts in this book, but my own need to perform easy lifts produced the adjustments to other standard lifts which are made easier by even the most minimal patient participation. As a physiotherapist I have been concerned in translating the word *care* into action by the patient as well as by my students and myself. Care to me means compassion which translates to doing something about caring. So patient involvement in lifting grew as did my demonstrations for the Multiple Sclerosis Societies in the United Kingdom and abroad. Always I am asked for a book with all the teaching points to enhance the film *Moving and Lifting the Disabled Patient.*

A fortuitous participation with Dr David Stubbs and his research team in a conference on back pain led to my own discovery that my work was more widely recognized than I realized and at the same time the details were less widely known. From that meeting this book grew. The researchers in the Department of Human Biology and Health at Surrey University invited me to have my lecture/demonstration video-taped and my methods recorded using the 'radio pill'. I then realized the finer teaching points I have used to prevent back injury for 30 years should be put on detailed record before I become too incapacitated by age and disease to demonstrate them myself.

My very grateful thanks are due to Professor P. R. Davis, Department of Human Biology and Health, for his interest and for kindly agreeing to write the Foreword, and to both Dr D. A. Stubbs of the same department at Surrey University and Dr J. D. G. Troup, Honorary Consultant Physician, Department of Rheumatology, Royal Free Hospital, and Senior Research Fellow, University of Liverpool for encouraging and advising me. I am very grateful to Dr D. A. Stubbs for Figure 35 recorded in the Materials Handling Research Unit of the University of Surrey.

I also owe much gratitude to Mrs Anne Raistrick, SRN, OHNC, Occupational Health Nursing Officer, Dewsbury, and Mr Christopher Dickenson, SRN, Tutor at the School of Nursing, Bradford, who read the manuscript and advised on the finer points of nursing, and to Mrs Barbara Turner, MCSP, DipTP, for criticising the manuscript.

Mrs Gill Robinson, MCSP, DipTP, read the manuscript at many stages and advised and helped with the photography as did Mrs Wyn Stokes, Mr John Rook, Mr Andrew Spink, Miss Jane Garrity and Mr Richard Stephenson. The photographer Mr Peter Harrison, AIMBI, has expertly translated my ideas to his medium and I am grateful for his help and advice, and to Bradford District Health Authority for the provision and use of facilities for the photographic sessions.

My thanks are also due to many hundreds of people involved in nursing and lifting and from all specialities who have attended courses and encouraged me to undertake this task. I am grateful to Mrs Muriel Cawthra who translated almost indecipherable scribble into a readable manuscript.

The artistic and editorial staffs of Blackwell Scientific Publications have interpreted my ideas and hopes for the book and by constant encouragement have helped this book to be the reality for which I wished.

Bradford, 1981   *Margaret Hollis*

PART 1   INTRODUCTION

Chapter 1   Friction, Posture, Bracing and Commands

Lifting and handling in relation to sick people can be a most stressful task and the ongoing research is revealing that there is a distinct relationship between the worker with the sick and disabled and the worker in industry. Both lift heavy loads which require approach, preparation, lift and lower. But there are two major differences:

a crate load cannot be asked to help
a human being can be hurt

In both cases the lifter can also be hurt.

FRICTION

Friction is the force which tries to stop one object sliding on another and is of use if you wish to keep something in position, e.g., a pair of nonslip shoe soles on an unpolished floor.

The objective of any lift done with and for a patient is to change his or her position or place with ease and comfort to lifted and lifter. However, the patient may be subject to two major hazards.

Frictional Effect   If frictional effect (coefficient of friction) is low then the two objects will slide on one another and a hazard may arise, e.g., when patients slip down in a bed or chair because they are wearing artificial fibre clothing. Friction between the patients' skin and the cover on which they are resting and along which they may be dragged is most usually a bottom sheet on a bed and is a serious hazard.

Patients with thin or elderly skin which has reduced elasticity on which they may have been lying for some time are especially vulnerable to the grazing effects of friction in lifting. This is more so if the friction is applied to bony prominences such as the sacrum, elbows, the malleoli and epicondyles.

Tearing Effects at Vulnerable Joints

The shoulder is the most often damaged especially when it is unprotected by normal muscle tone. If a 'drag' lift is performed by putting the radial border of your forearm under the patient's axilla from in front or behind and a 'dead weight' lift is done, then the shoulder joint is likely to be damaged in two ways. There will be agonizing discomfort to the patient at the moment of lift and unnecessary and possibly damaging pressure to the contents of the axilla (artery, vein, nerves). There will be trauma to the joint structures as the patient's weight is suspended (Fig. 1) by his or her shoulder joints, thus forcing the joint surfaces apart. The deep shoulder muscles in the normal joint may offer some protection, but will be unable to do so if the arm muscles are weak or paralysed as in stroke-disabled patients. Minor trauma will occur if hard bony grasps are used by the lifter which will cause minor bruising or tearing sensations of the skin on the subcutaneous tissues.

One major reason why patients 'fight' a lift is that they are being hurt. Another reason is lack of information. All patients should be told what to do, even if it is 'do nothing, keep still' so that they can participate and not be frightened into unwanted, but to them protective, movements.

Damage to the lifter occurs for reasons not yet known but the following are the most

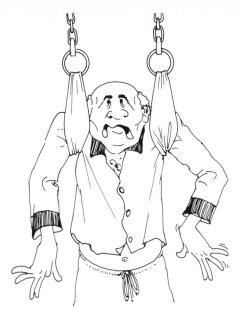

Fig. 1. The patient is suspended by the shoulder joints.

likely factors to contribute to the lifter's inability to observe safe procedures.

Too heavy a load.

Too heavy a load held out of balance for too long.

Standing too far away from the load.

Poor access to the patient so that the posture of the lifter is dictated by obstructions.

Too tight and unsuitable clothing worn by the lifter so that movements of the thighs are restricted either by tight skirts or by modesty preventing free and suitable movements of the lifter's body.

It is recognized that we do not yet know the causes of back and neck injuries in lifting. The

5   *Friction, Posture, Bracing and Commands*

realistic estimate that nearly half the nursing profession suffers from back pain indicates the need for improved techniques with more practice on the normal person to gain skill and understanding of the principles involved, not only in lifting but in dealing with the human being.

The application of safe limits is not easy, but one important rule is easy to apply: IF YOU CANNOT MOVE THE PATIENT DON'T GO ON TRYING—GET HELP. Even unskilled help will provide someone to share the load and a trained lifter can easily instruct a second helper.

CORRECT POSTURE

The next most important action of a lifter after getting help is to remember to support his or her own back correctly and to use the correct movements and muscle actions of the limbs.

The leg muscles are the strongest in the body; because they spend their lives pushing hard to propel your body weight in walking, they are built for a load. They should be given the opportunity to work over bent hips, knees and ankles by your allowing movements to take place there. Thus before lifting, start by bending these joints and straighten them during the lift. Equally, when lifting down, as in lowering a patient to a chair (see p. 69, Shoulder Lift for Transit) these joints should start straight and bend during the lowering action. It is just as important to keep your leg movements fluid by transferring your weight from one leg to the other as you lift. The arms should also be considered. An initially bent elbow can aid a lift by thrusting to a straight position (see p. 61) or bent elbows can be more bent to bring a load nearer to the lifter's chest and therefore over the base

Friction, Posture, Bracing and Commands

(see p. 94). However, when lifting individual body parts the elbows must not be bent and straightened to move the part. Instead your back must be moved first, then your arms, or you will be using your back as an incompetent crane (see Fig. 13).

The vertebral column should be bent as little as possible while you are lifting or carrying a load. If incompatibility of heights means some bending then the upper back should be bent a little but it is preferable to use the legs and squat (see Figs. 65a and 66a). The lower back must be braced and protected from potential damage.

BRACING

Bracing is an essential prerequisite of maintaining correct posture during a lift. There are two aspects to bracing:

1. Bracing the lower back is done by an action called *dynamic abdominal bracing* in which the lower abdomen is contracted with a pull upwards and towards the side of the waist. The lower abdomen gets flatter but the waist gets bigger at the sides. Done correctly this action does not prevent normal breathing and does not raise the blood pressure and heart rate. The action is simulated in popular 'girdle' adverts (Fig. 2).

2. Bracing of the rest of the bodies involved in the lift may be taking up the slack in your own body and in the patient's. Bracing your own body will be achieved by slightly raising your head and elongating your back as you do the dynamic abdominal brace. Your body should feel firm but not stiff and rigid. Bracing the patient's body will most often mean taking up the slack in his or her body as you do so in your own, but there is one special sort of bracing which must be considered.

7   *Friction, Posture, Bracing and Commands*

In many of the lifts described the patient's shoulder girdle is fixed and prevented from being elevated by the nature of the grasps prescribed; either the patient pushes his or her own arm down against the lifter's grasp or the lift is made through the lower end of the humerus and against the fixed upper end of

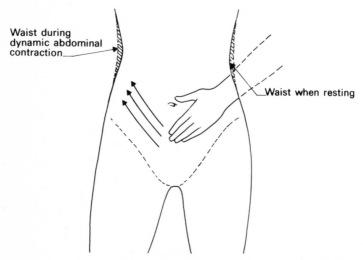

Waist during dynamic abdominal contraction

Waist when resting

Fig. 2. The lower abdomen is braced upwards and with a pull towards the waist which becomes wider at the sides.

the bone. What should not be allowed to occur is compression of the patient's soft tissues and an upward drag on the shoulders. This is what happens in the through arm lift if the lifter grasps his or her own hands across the patient's breast tissues instead of grasping the patient's forearms.

8   *Friction, Posture, Bracing and Commands*

COMMANDS

ONE PERSON SHOULD ALWAYS BE IN
CHARGE AND ISSUE THE COMMANDS
DURING A LIFT (Fig. 3). Remember that two
people are always present—the patient and
the lifter and the latter should take charge in
this case. If more than one lifter is used it is
traditional for the person in charge to lift at
the patient's head end. It is preferable to

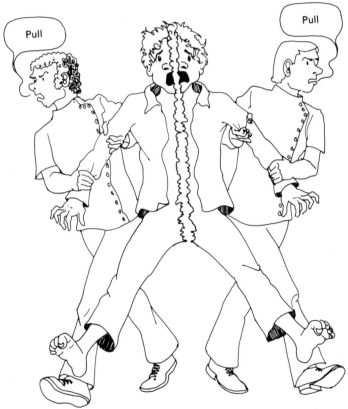

Fig. 3. One person should give the commands.

ignore this rule in favour of the most skilled or strongest or tallest person taking the appropriate part and to allocate the command if necessary to the person with the best line of vision of the setup.

Words not *counts* should be used as they convey meaning all the time and do not have to be constantly retaught to a lifting team. A count may consist of

Ready One Two
Ready One Two Three
One Two
or
One Two Three

or any other permutation.

How do the patient and other lifters know when the count has stopped and lifting must start? Word commands should consist of a sequence in which

READY indicates action is imminent.

HEADS UP may be for lifted and lifters.

BRACE means take the strain and do a correct dynamic abdominal brace (see p. 7).

LIFT, LOWER, and GRASP, PUSH, PULL, WALK, TURN and LET GO mean precisely what they say.

MAKE A TIDY PARCEL

The plea 'Make a tidy parcel' from the postal authorities is one which should also be a lifter's slogan (Fig. 4). The patient should not be lifted until the limbs are positioned and possibly are either tucked up or under the control of the lifter(s). If the arms are taken towards the midline of the body then the patient can push them down the thighs to help in sitting up (Fig. 5); if they are crossed they may be grasped by the lifter to form a

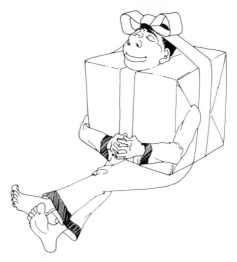

Fig. 4. 'Make a tidy parcel' of the patient.

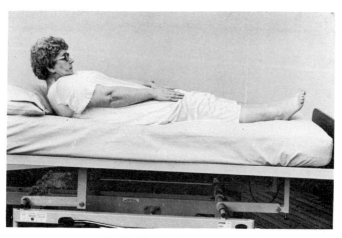

Fig. 5. The hands are pushed down the thighs as the head is raised.

shorter lever (elbow and through arm lifts—see Figs 25, 26, and 28a and b). The patient's legs and feet should be placed side by side so that both feet may be stopped from slipping (see Fig. 75) and the knees may be blocked (see Fig. 75) by one of your knees.

An untidy parcel is not only a patient whose arms are outside the outline of the trunk and whose legs are apart in sitting or lying but also could be a patient whose arms are waving about above both your heads when you are standing. Such a position increases instability and may cause loss of balance and a fall.

Chapter 2   Mechanics of Stability and Stances

MECHANICS OF STABILITY

Comprehension of the mechanical principles on which stability and equilibrium are founded is necessary to understand the importance of correct placing of everyone's feet during a lift.

Base   Base is the area on which an object is supported, the larger the safer if support is all that is needed. It includes the area between the feet so that the base of a bed is 91 cm × 198 cm (3 ft × 6 ft 6 in), that is, the area between the bed legs.

The base of a person standing consists of his or her feet and the area between them (Fig. 6); thus a base will be larger and safer when the feet are apart because it offers more

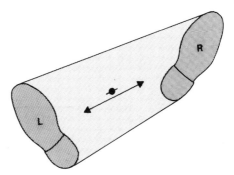

Fig. 6. The base; the *whole* of the shaded area is the area of support.

stability. The base in Figure 6 allows the body weight to be transferred from foot to foot but still offers stability. Any position of the feet in which they are apart not only offers stability but allows your body weight to be transferred from one foot to the other as you move a load.

Centre of Gravity

Centre of gravity is the point in the body at which all forces operate. In the standing human body it lies at the level of the second sacral vertebra, the bony prominence at the top of the buttock cleft, and light pressure applied at this point will cause the body to move forward (Fig. 7). Try it on a colleague. Then ask him or her to resist. You should be able to push your colleague easily even

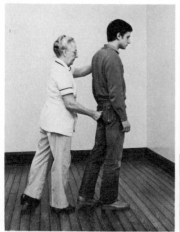

Fig. 7. *(left)* The centre of gravity of the upright human body is at the second sacral vertebra. Push there and the patient must move forwards.

Fig. 8. *(right)* A push at the same level on the side of the pelvis will cause side-stepping.

though you use only one finger tip. Next try standing alongside a colleague and gently but swiftly push on the side of the buttock at the same level. A sideways movement of the body will follow and he or she will side step away from you. You may need to offer protection to prevent a fall by putting your hand round to hold the waist at the opposite side (Fig. 8).

The centre of gravity is a key point at which pressure must be exerted to *help* a weak patient to become upright. It is a 'locking' point at which pressure must be exerted to keep a patient upright (Figs 78a and b), and most people will require counterbalancing pressures in the opposite direction *above*, across the shoulders, and *below*, across the knees from the front; otherwise the patient will always move forward.

The individual parts of the body each also have a centre of gravity and it lies at about the junction of the upper and middle third of each part. This is the level at which each part of a limb should be supported to move it most easily (see p. 53 and Figs 45–49).

The Line of Gravity

The line of gravity is a vertical line which falls through the centre of gravity and within the base (Fig. 9). If the line of gravity falls outside the base, balance is lost (Fig. 10). Thus, although the base (as illustrated in Fig. 11) is wide and safe for staying still if more forward movement of the patient's body occurs, the base will not be safe as the line (and centre) of gravity fall outside the base. Too wide a base allowing too large a range of movement in transferring weight from one foot to another may cause instability and loss of balance when a lot of weight gets too near the edge of the base. For example, where the lifter is in lunge standing (see Fig. 54) the movement of the patient must be within the length of the distance between the lifter's feet.

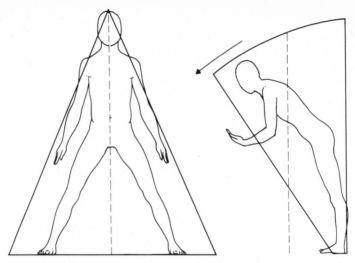

Fig. 9. *(left)* A large base with the line of gravity falling within it is very stable.

Fig. 10. *(right)* A small base offers poor stability as the line of gravity falls outside it.

A common error is to give a big heave and swing when lifting, so that the patient is moved a long way at each lift. This is a major mistake. The greatest truncal stresses in every lift occur in sustained loading, and, in addition, the lifter has to balance both the patient's and his or her own weight at the edge of the base. This also is very likely to cause loaded flexion and rotation of the vertebral column which is more likely to cause damage.

In Figure 12, the lifters are forming a bridge with their legs as the pillars and their arms as the span. The patient rests on their forearms and hands and the base is formed by their feet and the area between. Although this base is wide, it is not stable unless the two lifters lean back simultaneously as they take

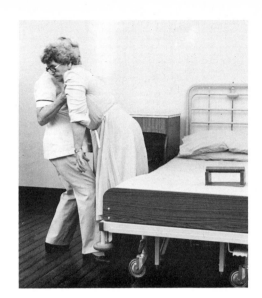

Fig. 11. Although the base is wide and in the direction of movement, if the patient leans further forward the line of gravity will fall outside the base and both people will lose their balance.

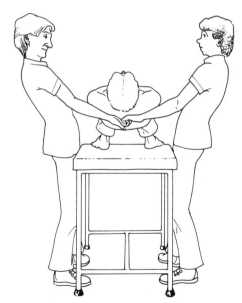

Fig. 12. The lifters form a bridge with their arms as the span and their bodies and legs as the pillars. If both lean back the load will move upwards.

Fig. 13. An incompetent crane.

the load and prior to moving it up the bed. If one lifter does less work than the other then the base for the lifter making the greatest effort becomes only his or her own feet and the area between, as half the span is lost. Stability is then lost and, in effect, that lifter becomes an incompetent crane (Fig. 13).

STANCES FOR LIFTING AND SUPPORTING

To conform to the mechanical rules outlined in the preceding paragraphs, it is necessary to consider the positions in which lifters should stand to maintain a safe base for themselves and their patient and to maintain good balance during the lift and transfer of weight.

NEVER LIFT WHEN THE WHOLE OF BOTH FEET ARE CLOSE TOGETHER.

The commonest foot positions are:

Walk When one foot is in front of the other as in taking a step and weight can be transferred forwards and backwards. This stance (Fig. 14) is used for the through arm lift up a chair, the forearm grasp lift and may be used for the waistband and belt lifts. It is an optional stance with Stride when supporting individual body parts.

Fig. 14. Walk standing.

Lunge   When the feet are apart and the toes point in directions at right angles to one another. The knees and hips should be flexible and it should be possible to transfer body weight from one foot to the other by bending first one knee while the other straightens and vice versa. This stance has two variants:

1. When alongside a bed doing the shoulder and orthodox lifts, the feet are hip width apart and along the line of the bed one foot pointing under the bed and the other in the direction of movement (Fig. 15).

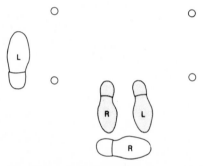

Fig. 15. Lunge standing beside a chair. The chair legs (circles) indicate positioning of patient's feet (shaded) and lifters feet (unshaded)

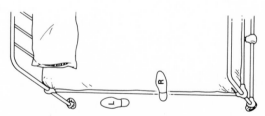

Fig. 16. Lunge standing beside a bed.

Mechanics of Stability and Stance

2. When alongside a chair doing the axillary and elbow lifts, one foot is put across the front of the patient's feet and the other foot is along the side of the chair and pointing backwards (Fig. 16).

Stride When the feet are apart about the width of one's own hips and weight can be transferred sideways from one to the other. This stance is used optionally with Walk when supporting individual body parts. It is also the stance used when you wish to squat stand (Fig. 17).

Fig. 17. Stride standing.

Ten-to-two When the heels are together and the toes apart. The knees should be flexible and bent outwards. This stance (Fig. 18) is used for the waistband and buttock lifts. It is the least stable of the foot positions and once the patient is standing satisfactorily, one foot should be moved back to the walk position.

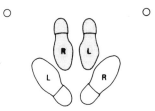

Fig. 18. Ten-to-two standing. The lifter's feet (unshaded) placed facing the patient (shaded feet) seated in a chair.

Chapter 3　The Grasps

Grasps for lifting should always be designed to be comfortable remembering that every grasp will also bear a load as the lift is performed. A grasp in which your hand becomes tense or drags on skin or body folds as the load is taken, will be painful for you or the patient or both. Always be aware of what you feel as you lift. If you feel a drag or hard pressure on a bony surface then the patient will feel it more.

The human hand can remain soft and relaxed while lifting and should always be kept in contact by means of the palmar surface. The forearm should *never* be used as it often is by the thoughtless in a drag lift. The two most painful grasps are: the forearm put with the bony edge of the radius uppermost under the patient's axilla to drag him or her; and the knuckles of the hands digging in under the lower thoracic spine during the orthodox lift.

There are three grasps which lifters use on one another: the finger grasp; the double grasp; and the single wrist grasp.

The Finger Grasp　The finger grasp is one in which the two lifters hook their fingers on each other's (Fig. 19). It is a useful grasp when lifting an obese patient as it gives the arms extra reach.

The Double Grasp　The double grasp on wrist, forearm or elbow is one in which each lifter grasps the wrist, forearm or elbow of the other (Figs 20 and 21). The wrist grasp requires one lifter to turn

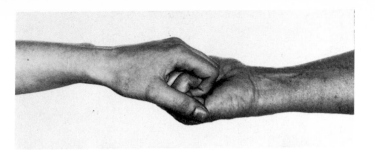

Fig. 19. The finger grasp gives a long reach.

Fig. 20. The double wrist grasp.

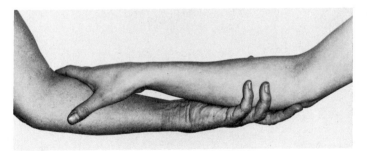

Fig. 21. The double forearm grasp is comfortable.

The Grasps

the palm fully up (supination) and the other lifter to turn the palm fully down (pronation). If either lifter cannot achieve one of these positions then this grasp is likely to cause the knuckles of the uppermost hand to dig into and hurt the patient and the grasp should not be used, but you should use the double forearm or elbow grasp instead.

Fig. 22. The single wrist grasp used when one person has a weak grasp.

The Single Wrist Grasp

The single wrist grasp is one in which one lifter grasps the wrist of the other whose hand rests palm upwards (Fig. 22). This grasp should be chosen when full range supination or pronation cannot be achieved or when one lifter has a weak grasp.

There are eight grasps which lifters use on patients: the relaxed hand grasp; the through arm grasp; the axillary grasp; the elbow grasp; the waistband grasp; the buttock grasp; the forearm grasp; and the palm-to-palm thumb grasp.

The Grasps

The Relaxed Hand Grasp

The relaxed hand grasp is one in which the hand, palm uppermost, conforms to the contour of the part to be lifted. The natural rest position of the human hand is with the fingers and thumb a little apart and very slightly flexed at each joint, and it can easily be adjusted to allow a body part of any size to be supported. This is the grasp which is used in lifting or supporting individual limbs or parts (Fig. 23).

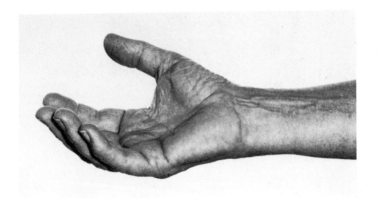

Fig. 23. The relaxed hand grasp.

The Through Arm Grasp

The through arm grasp is one in which the lifter assists the patient out of a slumped and into an erect position by first pushing carefully on the back of the patient's head or shoulders. The lifter's hands are then inserted, from behind, between the patient's chest and upper arms to grasp the patient's forearms (Fig. 24).

The Grasps

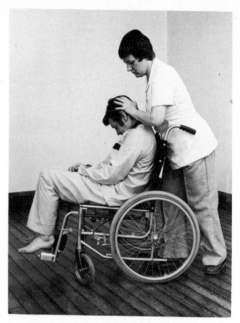

Fig. 24. The relaxed hand grasp is used to push on the back of the patient's head to prepare him or her for the through arm grasp.

Either

The patient grasps one wrist with the other hand and the lifter grasps at about the middle of the forearms (Fig. 25)

or

the patient who is incapable of grasping the wrist has both forearms tucked across one another and the lifter then grasps the forearms nearer to the elbows (Fig. 26).

This grasp allows the lift to be transmitted to the patient via their upper arms. The patient should never be grasped across the chest which is a common error as it is agony on breast tissue.

26   *The Grasps*

Fig. 25. The through arm grasp. The patient grasps one wrist with the other hand.

Fig. 26. The through arm grasp. As the patient cannot grasp, his or her arms are tucked across the chest and the lifter holds the forearms as near to the elbows as possible.

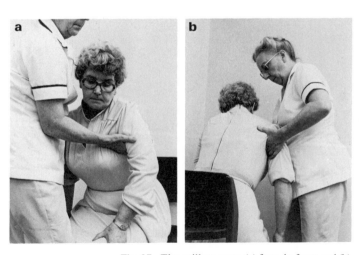

Fig. 27. The axillary grasp (a) from in front and (b) from behind.

The Axillary Grasp In the axillary grasp the lifter stands obliquely at the side of the patient. Your forward foot is across the feet with your knee of the same leg across the knees, and your other foot is placed at right angles beside the chair (see Lunge Stance, p. 20). The patient is leaned forward and you put your forward hand, bent at the knuckles and thumb out, so that your palm and fingers are under the axilla of the opposite side from in front. Your rear hand is put in a similar position under the nearside axilla from the rear (Figs 27a and b).

The Elbow Grasp The elbow grasp is one in which the lifter again stands with the forward foot blocking the patient's feet and the knee of the same leg blocking the knees. Your other foot is at right angles at the side of the chair. The patient is leaned forward so that the nearside shoulder rests on the front of your trunk. Lean across

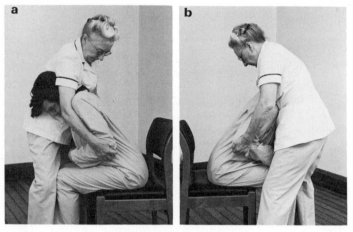

Fig. 28. The elbow grasp. (a) The front hand and (b) the rear hand.

The Grasps

to reach with your front arm over the back of the patient's neck, in front of the opposite shoulder and grasp under the farside elbow (Fig. 28a). The same grasp is taken under the nearside elbow (Fig. 28b). Both elbows are held tucked in to the patient's waist and a little backwards from the mid-axillary line.

The Waistband Grasp

In the waistband grasp the lifter

either
Stands in front of the patient, one foot forward between the separated feet. Both of you lean forward so that your heads are over each other's shoulders. Insert your thumbs into the side of the patient's trousers and grasp ready to lift upwards and forwards.

or
Stands in front of the patient, heels together, toes apart (see Ten-to-two stance, p. 21) so as to block the feet. Bend your knees on each side of the patient's knees to block them. Both of you lean forward, putting your heads over each other's shoulders. Grasp the waistband and the lift is performed (Fig. 29).

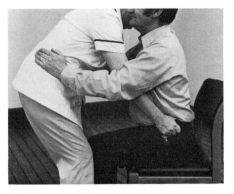

Fig. 29. The waistband grasp.

The Grasps

There are alternative hand positions for the patient.

a. The hands may push on the knees (as in the axillary lift, Fig. 25a).

b. The hands grasp or rest on the lifter's waistband or belt.

c. The arms may rest on the lifter's shoulders.

When the patient is nearly upright the lifter should first transfer one hand to the sacrum and push the pelvis forward and then the other hand to between the shoulder blades to hold the patient securely (see Fig. 91).

The Buttock Grasp

The buttock grasp is one in which the patient who is not wearing a waistband may be lifted as in the preceding lift, but the lifter's hands pass between the patient's chest and upper arm and then under the buttocks from the side (Fig. 30). The procedure then followed is the same as in the waistband grasp.

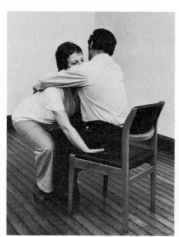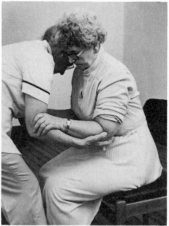

Fig. 30. *(left)* The buttock grasp used to lift patients with no waistband.

Fig. 31. *(right)* The forearm grasp.

The Forearm Grasp

In the forearm grasp the lifter stands in front of the patient, turns both palms up and grasps the patient under the elbows. If possible the patient should grasp the lifter's elbows (Fig. 31). This grasp is used to help a patient to stand up from sitting or to keep his or her balance whilst standing.

The Palm-to-Palm Thumb Grasp

In the palm-to-palm thumb grasp the lifter, standing to the side of the patient and with the outer foot forward as in walking, presents an upturned palm, thumb out, to the patient who

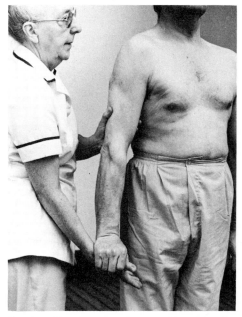

Fig. 32. The palm-to-palm thumb grasp. The patient grasps; the lifter does not. The lifter's other hand pushes on the back of the arm just above the elbow to help straighten it. Note that this grasp allows the patient to use the lifter as a support for walking.

The Grasps

rests his or her palm on your palm and grips your thumb. As this grasp is used when helping a patient to stand or walk the patient may need additional assistance from your other hand. You either use the relaxed hand grasp (see p. 25) to apply forward pressure above the elbow (Fig. 32) or give greater support by putting your arm across the patient's back to hold the waist at the opposite side.

Chapter 4   Mechanics and Performance of Rocking Manoeuvres, and the Use of Blocks

Patients should not be lifted, pulled or pushed in any sense of the word if they are capable of moving themselves. A few minutes occupied in teaching a patient how to move are well spent for the patient's progress and pleasure and for the sake of reducing the daily work load on the lifter, which may be as high as 1874 kg per day for a nurse.

To this end the ways in which we move ourselves, especially small distances when friction tends to resist the movement, need consideration.

Consider how many things rock and move—those weighted toys which rock back and forth and sometimes turn right over, cradles, chairs, swings and people. The use of rocking chairs is now being advocated to reduce low back pain, but use of the rocking principle can prevent its onset.

Oscillation   Oscillation of a body is its swinging back and forth. A clock pendulum oscillates, so does a garden swing. When individuals use repeated movement they usually do so to put more energy into the effort previously made; each oscillation of the body being made bigger by the additional effort. Think of when you used a garden swing. You sat down and walked your feet back, let go and swung forwards. As you swung back again you pushed your feet against the ground thus pushing yourself and the swing higher. After a few oscillations you leaned your trunk back and straightened your legs at one end of the arc of movement

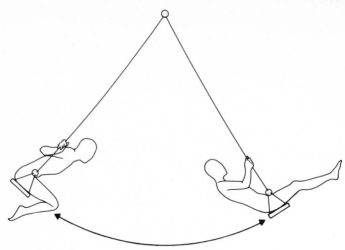

Fig. 33. Adding energy to each oscillation causes the swing to go higher.

and leaned your trunk forwards and bent your legs at the other end of the arc of movement. If you timed your leg and body movements correctly you added energy to each successive oscillation and the swing went higher. Eventually you could stop these actions, therefore reducing your effort, and continue swinging for a time (Fig. 33). Note that a swing rocks from the hooks above the movement.

If this oscillatory activity can be used jointly by patient and lifter then the effort for each will be less, and greater actions can occur for minimal effort. This is especially so in standing up from the sitting position. A lifter can use forward rocking with a patient and by the use of correct timing of his or her help can greatly assist the patient. Note that in this case the swing occurs from the knees, and the *upper* end of the body will perform the arc of movement (see Fig. 34).

Timing Timing is the regulation of the speed with which something is performed so as to produce the most effective result. To use again the analogy of the swing: if the leg and body movements are performed too early or too late during the arc of movement they will add less effective energy to the movement of the swing as they will be against the movement of the swing at that moment. Timing is important in lifting for three reasons.

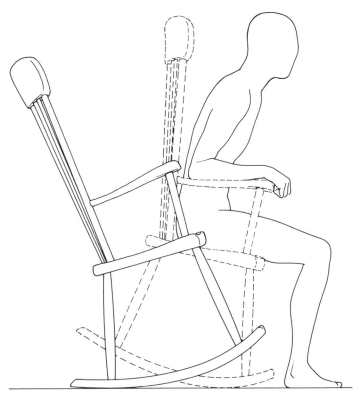

Fig. 34. The body swings up from the knees when repeated attempts are made to stand up.

Rocking Manoeuvres

1. The lifter working with the patient must allow the patient to initiate the movement and then add assistance at the right moment so that the movement is aided and does not start as a drag on the patient.

2. More than one lifter working on the same lift must exactly time not only the speed of the effort but the quantity so that the team move both in unison and an equal amount, and the patient is then lifted smoothly.

3. In either of the preceding cases the third reason must be used. That is, the operative command(s) for the action must be said and if necessary repeated in correct time with the actions of the patient and the lifter. For example, to say PUSH as the patient sits down is too early in his or her attempt to rise, and to say it after he or she has started to get up is too late. The correct time is at the precise moment *after* the patient sits down and *before* he or she starts to rise again.

Inertia

Inertia is the property of a body by which it opposes any change in its state of rest or motion. A body possessing inertia of motion finds it easier to stay in motion. Inertia of motion is of use in moving a patient if the movement is initiated by the patient. The lifter finds it easier to keep that movement going by adding a lot less energy than if he or she has to initiate the movement. It is easier to push a car that is already moving than to start it moving.

ROCKING THE HUMAN BODY

The human body can rock

either
Forwards and backwards
1. moving the head most
2. moving the pelvis most

Rocking Manoeuvres

or

From side to side (lateral rocking)

Rocking in either direction is of use to the sitting patient.

Forwards and Backwards Rocking

Moving the Head

Rocking forwards and backwards is more commonly done from the hips as in rocking with laughter or grief and is a primitive movement often used by the mentally handicapped as well as the normal patient. However, to rock the upper part of the body to aid its movement the rock must take place from the knees when the feet are fixed on the floor and are immediately under the knees. Observe weak or elderly persons attempting to get out of a fairly low armchair. They almost invariably have more than one 'go' and rock the trunk forth and back to gain a greater swing up with each effort until, at last they are able to rise. If such a rocking action is used by a patient and assisted by a lifter, both will do less work and the lifter will have considerably less load on the back. Timing to work together is all important, but it is possible for a lifter alone to rock an incapacitated patient forth and back, build up the size of each oscillation until eventually the patient is lifted or moved at the peak of the highest oscillation.

There are two ways of using forwards and backwards rocking of this type:

without patient participation and

with patient participation.

WITHOUT PATIENT PARTICIPATION

Forwards and backwards rocking without patient participation is a technique used in the elbow lift, and the loading is recorded on Figure 35. This is a graph of a recording of

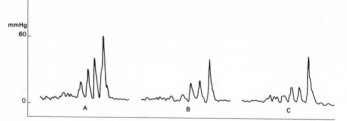

Fig. 35. A graph recording truncal, intra-abdominal, pressures during performance of the elbow lift of a seated person. A represents the first attempt, B and C subsequent attempts after the lifter was retaught. Note increased skill lowered not only the pressure during the actual lift but very greatly lowered the pressure during the preparative rocking phases (from Materials Handling Research Unit, University of Surrey).

abdominal (truncal) pressures using a radio 'pill' swallowed by an inexperienced male lifter performing this lift. Reteaching and improvement of skill were undertaken at each attempt of which *A* was the first and *B* and *C* the subsequent attempts. The highest peaks show the truncal stresses during the actual performance of the lift and the preceding lower peaks show the truncal stresses every time the patient was rocked forwards and partly out of the chair. Note that the major peak of stress was highest at the first attempt, *A*, and considerably lower, though at about the same level, on attempts *B* and *C*. But the interesting feature of this record is the dramatic reduction in the peaks during the preparative rocking phases, when each attempt shows greater skill with greater use of *all* the lifter's body. Initially on attempt *A* the lifter was using a lifting action, but on reinstruction he started to use transference of his weight from one foot to the other to rock back and forth from his feet as he rocked the

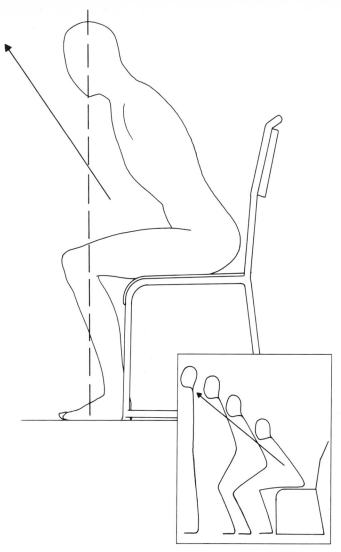

Fig. 36. When preparing a patient for standing up, first align the head above the knees and feet so that the head can lead the movement.

patient forwards and back and the 'lift' became less of a load on his lower back and a bigger oscillatory movement.

The mode of performance of this rock is for the patient to be seated near the front of the chair if possible and to be leaned forward so that the head is in line with the knees and the feet which should if possible be at right angles (Fig. 36). The lifter takes up the Lunge stance (see p. 20) with one foot to the side of the chair and the other foot across the front of the patient's feet. The knee of this leg is bent across the front of the patient's knees. Using the elbow grasp the lifter braces, then rocks back onto the rear foot, i.e., that which is across the front of the patient's feet. The lifter then rocks forwards onto the foot at the side of the chair and, maintaining the grasp, continues to rock a very little distance further back each time until the patient's bottom leaves the chair seat to a sufficient height to allow the patient to be swung to another position.

WITH PATIENT
PARTICIPATION

Forwards and backwards rocking with patient participation is the technique used in the axillary lift and was originally devised for stroke-disabled patients who use their sound arm and leg to help themselves. In this rock the patient 'leads' the movements and the lifter assists by adding energy to the patient's effort. In consequence, skill on the part of the operator is all important in performing this rock with the patient.

The mode of performance of this rock is initially exactly the same for the patient's position as for the previous rock except that the sound hand rests on the sound knee as though to grasp it and the patient is taught to push on that hand and feel the effect (the bottom should start to ease off the seat); to push on that foot and feel the effect (the bottom raise is enhanced). The lifter takes up the same position as in the preceding description *but* keeps her or his weight mainly on the foot at the side of the chair, also keeping the body to the side. Then use the axillary grasp and the axillary lift (see Figs 27a and b, and 75).

The skill in using this rock lies in:

1. The combined effort of patient and lifter being timed exactly together by the lifter, who adapts to the speed of the patient's movements.
2. The lifter recognizing the need for the patient to push first in order to straighten the pushing arm and then, as each successive rise of the patient from the chair brings him or her nearer to the erect position, to 'let go' of the thigh so that he or she may be assisted to stand up.

Moving the Pelvis   The second forwards and backwards rocking manoeuvre involves moving the pelvis. If the arms are fixed the patient can swing the body through them as in attempting to move up or down a bed. However, to swing the pelvis in this way and reduce friction the body must be lifted well clear of the seat whether it is a bed or chair. The arms must be pushed straight down beside the body with the elbows fully extended. Most people then discover that their arms are too short to stretch down beside their body and lift it off their seat. This is more especially so when the seat is made of compressible material such as bed springs or upholstery foam. Endeavouring to 'lengthen' the arm by using the knuckles only diminishes the area with which the push is made and condenses the load on to a smaller area of springs or foam which compress even more. The force for lifting just causes more compression of the seat and the buttocks are not raised. To help this movement every patient should have the chance to

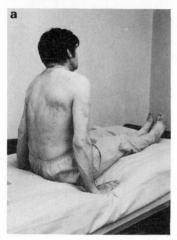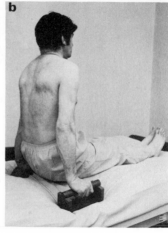

Fig. 37(a). A patient unsuccessfully attempting to lift his buttocks clear of the bed and (b) the same patient using hand blocks to spread the load is now successful.

use some form of hand block (see p. 47) when moving about, especially on a compressible surface. Figure 37*a* and *b* shows the difference between the endeavours of a patient with and without hand blocks (see also Fig. 44). Such a lift can be accompanied by a swing of the buttocks backwards or forwards and is friction free.

Lateral Rocking Sideways rocking can be used to obtain forwards or, less frequently, backwards motion of the body so that a patient may be moved to the edge of a chair, back into a chair or up or down a bed. The trunk must be leaned forwards with the arms in front of it and the head in line above the knees and feet which are pulled well back. A rocking movement from side to side is initiated.

Rocking Manoeuvres

Either

The pelvis will rotate forwards alternately with the rock, i.e., as the head and shoulders rock to the right the left side of the pelvis will move forwards and nearer to the chair front (Fig. 38). Clothing of high frictional resistance on a seat of high frictional resistance may impede but will rarely halt this movement. The forward movement of the pelvis is thus an inadvertent action by the patient which does not involve deliberate lifting by either the patient or a lifter.

or

A similar action can be performed deliberately and more slowly when a patient is asked to lift first one thigh then the other forwards. This is especially used if a patient cannot lean

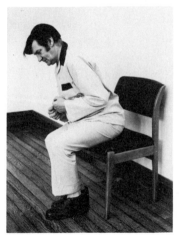

Fig. 38. *(left)* Sitting leaning forward, sideways rocking of the upper trunk to the right causes forward movement of the left thigh.

Fig. 39. *(right)* Sitting, unable to lean very far forward, the right shoulder and leg are hitched forward together so the pivot is on the left buttock.

very far forwards. When this rock is performed there is usually an automatic sideways lean away from the thigh to be moved putting greater weight on the buttock on the side of the lean. The patient pivots on this buttock and the whole of the non-weight-bearing side swings forwards (Fig. 39). This action is commonly seen in people with a lot of joint stiffness, e.g., rheumatoid arthritis.

If patients have difficulty with either of the above actions they can be told to use their arms on the chair arms. They can

either

Push on the chair arm to straighten the elbow on the side of the thigh to be moved forwards (Fig. 40)

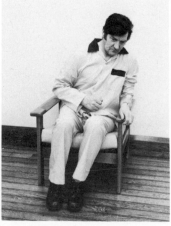

Fig. 40. *(left)* A push on the chair arm straightens the left elbow and allows the left thigh to be lifted forward.

Fig. 41. *(right)* A pull on the chair arm bends the left elbow and allows the right thigh to be lifted forward. These lifts should be used for a patient with only one useful arm.

or

Pull on the chair arm and bend the elbow to allow the opposite thigh to be moved (Fig. 41).

These two actions relieve weight from the buttocks and allow them to be moved free from frictional resistance. Patients with only one useful arm can thus, by alternately pushing and pulling on their chair arm, move about in their chair and rock to the edge in preparation for other manoeuvres.

Backwards movement into a chair or up a bed is achieved in a similar way, the patient lifting the buttock in the desired direction.

If patients cannot move themselves to the edge of the seat in any of the aforementioned ways they can be moved by the lifter as follows:

The lifter positions the patient with the feet together and as nearly under the knees as possible and stands in front of the patient in the 'ten-to-two' stance blocking the *feet only* (see Fig. 18, p. 21). The patient's knees should not be held. Using the relaxed hand grasp just behind each of the patient's shoulders the lifter gently pulls on *both* shoulders until the patient is leaning forwards with an erect back. The palms are then moved to the outer surface of each shoulder, but the fingers kept behind the shoulders. A sideways rock is initiated, built up a little, perhaps once or twice to each side, then as the patient is leaned to the left the lifter pulls forwards on the right shoulder and the patient should move forwards on that side (Fig. 42). The next rock to the right should be accompanied by a pull forwards on the left shoulder and so on until the patient has moved as far as necessary.

For a heavily disabled patient a different grasp may be used. The lifter adopts the stance shown in Figure 15 and described on page

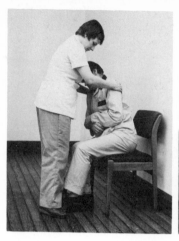

Fig. 42. *(left)* The patient is leaned forward and rocked from side to side so that pressure on alternate shoulders will automatically pivot the thighs forward. Note that the feet are carefully 'blocked'.

Fig. 43. *(right)* A heavily disabled patient has the sideways rock maintained and the lifter lifts alternate thighs forward. Note that the feet are carefully 'blocked'.

20, and pulls the patient into a forwards leaning position then puts one arm across the back of the shoulders to hold the opposite shoulder. Using this grasp to initiate a sideways rock, when the weight is off the patient's buttock on the side of the lifter's free arm she or he uses the relaxed palm grasp under that thigh to lift that side forwards (Fig. 43). The lifter maintains the rock, changes hand positions, and repeats as frequently as necessary. To allow time for a dexterous hand change the patient should be rocked

To right, left, right and lift the left leg;
To left, right, left and lift the right leg.

A rhythmical pendulum type of action

46   *Rocking Manoeuvres*

should be kept going, and will be assisted by an appropriate pivot of the lifter's body. Thus if your right arm is across the back of the patient's shoulders your pelvis should be pivoted towards your own right with more weight on your right leg. When rocking some patients you may need to side step to right and left alternately.

BLOCKS

Hand blocks need not be elaborate but can easily be made from a small pile of paperback books—a thickness of about 6–7 cm is sufficient—or a folded pad of newspapers or magazines and tied with string or tape (Fig. 44a).

The tape should not be tightly fastened as there must be room to slip the fingers under the tape on the top of the pile, the thumb remaining at the side of the books. Patients can then hold the book blocks and take them with them as they move.

Specially made blocks consist of wood and can be of two shapes: one to grip, a Handle Block; the other to rest on, the Palmar Block. Either of these blocks can be made by a handyman.

A Handle Block to grip consists of a wooden rod about 35 mm diameter and 220 mm long fixed to two end walls 90 mm

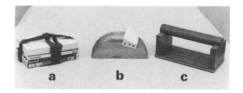

Fig. 44. The Blocks. (a) A pile of paperback books fastened with wide tape (b) the palmar block and (c) the handle block.

Rocking Manoeuvres

high × 70 mm wide × 20 mm deep which are attached to a base 220 mm long × 100 mm deep (Fig. 44b, see also Fig. 37b).

A Palmar Block to rest on is made of a wooden block 205 mm long × 90 mm wide × 70 mm deep and is shaped with a 50 mm length in the centre kept flat and the ends rounded as in Figure 44c. Velcro 120 mm long × 50 mm wide is fastened at an angle of 5° to the vertical to the side of the block and at the level of the end of one curve. The velcro should be fastened round the hand with the thumb out, so that it lies just across the knuckles and the palm rests on the flat part of the block.

Chapter 5   Selection of the Lift; Standard Procedure for a Lift; and Moving Body Parts

SELECTION OF THE LIFT

Selection of the appropriate lift for the job to be done is important both to the patient and to the lifter. To the patient it may mean that involvement in his or her own care permits arrival in a new position with maximum awareness of what is happening and with dignity and comfort. To the lifter it means that communication has gone on, teamwork has been involved, the manoeuvre has been happily and satisfactorily accomplished and he or she feels no immediate or long-term discomfort in consequence of the lift performed.

In some cases selecting the appropriate lift is so obvious that little choice is made, in others all the knowledge, skills and resourcefulness of the lifter may be called into action. The task will be made easier if the lifter has some foreknowledge of the patient and the condition from which he or she suffers.

The questions to be considered fall into several parts

Is the patient conscious or unconscious?

Are there any attachments which need special care such as monitors or I.V. lines?

Are there special problems such as flaccid or spastic limbs?

Are there full range movements at all joints?

Can the patient grasp, push or pull with the hands?

Is the patient blind or deaf?

Can the patient push with the full foot on the floor or the heel on the bed?

Can the patient be handled without causing pain?

Can the skin be handled at the necessary points of contact for that lift?

Where is the patient?

On a bed, chair, trolley or the floor?

What is his or her position? Lying tidily, untidily, on the back, side or front; slumped down or sideways.

Where do you want the patient to be eventually?

If you wish to take a patient from a slumped, sideways lying position to sitting in a wheelchair, a series of manoeuvres and lifts will be necessary.

The final question may be:

What does the patient weigh?

If the patient can help the weight may be less relevant than the question *how much can the patient help?*

So many people feel that to lift a patient into another position must be an instant procedure. In fact the instant lift is frequently one which can wait and is rarely imperative. The patient on the floor cannot fall further and making him or her comfortable and then getting enough help are the first priorities. The patient can be lifted off the floor when there is plenty of help and with greatest guarantee that he or she will arrive on a more comfortable support with minimum hazard to everyone—patient and lifters.

The choice of the appropriate lift to use will take all these factors into consideration but whichever is finally used it may be necessary to reposition the body parts first.

STANDARD PROCEDURE
FOR A LIFT

An experienced lifter should automatically follow a routine whereby the safety of the patient, the lifter (or lifters) and the equipment is ensured at all times, but until these reflexes are developed it is wise to practise a routine procedure.

Having said this, not every part of the procedure applies to every lift but experience soon increases skill to the point where lifters recognize what must be done for safety and effectiveness. The following routine is also that adopted for the descriptions of the lifts.

Decide on the Lift

Deciding on the lift is fully dealt with on pages 49–50; at this point it is necessary to ensure that, the decision having been made, all participants are told of the choice. This may just consist of two sentences: to the helper 'We'll use the shoulder lift' which presupposes that everyone in the lifting team knows the shoulder lift, and to the patient 'We are going to lift you up the bed and we want you to lean forward and then stay still'.

Position the Equipment

Positioning the equipment is more than a quick check on the brakes. It involves:
1. Removal of unnecessary and impeding linen and clothing to adjacent areas where it may be reached if necessary, once the patient has been moved.
2. The positioning of the new seat (chair, wheelchair etc.) with convenience and access.
3. Checking the patient's walking aids or appliances for preparedness in an adjacent situation.
4. Checking the height of the lifting surfaces.
5. Checking the stability and fixed situation (brakes or blocking) of bed and chairs.

Selection of the Lift

Inform and Position the Patient

As you position the equipment you should **simultaneously** inform and position the patient. This process may well involve preliminary movement of the patient to a more convenient position but at all times he or she should be told what is to be done. This may be a very simple sentence or a stream of information, for example, 'I'm going to lift you nearer to me, head first, cross your arms over your chest, now your legs, can you push on your heels and shoulders and help me to lift your tail? Now I'm going to cross this leg over that one, tuck both your arms across your chest, turn your head to look out of the door and over you go.' It may involve a series of small activities in some of which the patient, by participating is aiding his or her own recovery.

Position Yourself

Positioning yourself may be done **simultaneously** with the positioning of the patient, or it may need a series of separate manoeuvres. It should be done in advance of the instruction to the helper who may be taking up a similar position on the other side of the patient (shoulder and orthodox lifts). However, if the lifters are each to deal with a different body part (two-person through-arm lift, three-person lift) the helper (or helpers) should be instructed first.

Instruct and Position Your Helper(s)

Instructing and positioning your helper (or helpers) should be done partly by verbal explanation and partly by demonstration and example. The leader should check that the instructions are both understood and carried out. Grasps should be agreed on and, most important, the commands must be agreed on.

The Commands

In the commands—REMEMBER ONE PERSON MUST GIVE THE COMMANDS (see Fig. 3)—try to avoid using a count as there are innumerable variations on what the numbers mean. Use instead word commands.

READY tells everyone, patient and helper, that action is imminent. It should be followed by

BRACE tells the lifters to pull in the lower abdomen (see p. 7) and to increase the tension in their lifting muscles by taking up the slack in their positions. If the patient is to initiate the movement that command follows

either

PUSH

or

HEAD UP

LIFT follows immediately and is for the lifters so that they continue the movement initiated by the patient in response to the last command.

LOWER must not be forgotten as it is the command which starts to ensure that the lift will be completed successfully. It is often needed for transfer lifts.

Practice makes for better timing and a practised and skilled leader will lead comfortable and safe lifts.

MOVING BODY PARTS

Positioning of Body Parts

Movement of the whole body is most safely carried out if the head and limbs are safely positioned first. Use the relaxed hand grasp as it is the most comfortable. When you want to push your hand under a limb compress the chair seat or bed to allow your hand to pass under the patient's body. If this is done the patient's skin will not be dragged. It is easiest also to slide your hand under a body hollow and having got it far enough under to slide it up or down the limb or trunk to get it into the best position for the lift.

Supporting the Different Parts of the Body

The Head

Your fingers should be inserted from each side under the highest part of the neck and then slipped up to rest under the occiput and/or the chin (Fig. 45) If the head is face up the lift is made under the occiput. If the head is facing sideways the lift is made under the occiput and the lower jaw.

The Arms

Your fingers are

either
Inserted under the wrist and just above the elbow, then slipped up to the middle of the forearm and the upper arm (Fig. 46)

or
Slipped under the patient's downturned palm and below the elbow. The upper hand then slides up to rest under the elbow so that the bony prominence rests in the hollow of your palm. Your hand at the wrist should be spread out to support the patient's wrist and fingers (Fig. 47). This support is suitable for a flail (limp) arm.

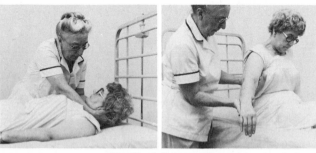

Fig. 45. *(left)* Supporting and moving the head.

Fig. 46. *(right)* Supporting the arm.

Selection of the Lift

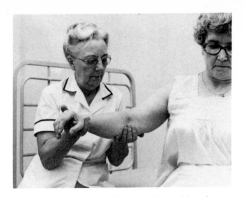

Fig. 47. Supporting the elbow, wrist and hand.

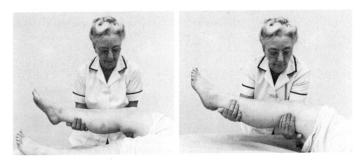

Fig. 48. *(left)* Supporting one leg.

Fig. 49. *(right)* Supporting both legs.

The Legs   Your fingers should be slipped under the knee and the ankle and then up the limb to rest just below midpoint of the thigh and the calf (Fig. 48). If both legs are to be lifted they may be placed side by side or crossed at the ankles and lifted together by sliding your hands under the furthest leg while supporting the nearest leg on the fully supinated lower forearm (Fig. 49). The legs then do not rest

          Selection of the Lift

on the bony edge of your radius with consequent discomfort to both patient and lifter.

Repositioning the Body Parts

When repositioning body parts either for comfort or in preparation for another manoeuvre ALWAYS LIFT BODY PARTS TOWARDS YOURSELF.

Move

1. either
The Head

or
The Head and Shoulders

this tells the patient where he or she is going (see Fig. 45).
2. The Arms which should usually be placed across the body unless the patient is to help.
3. The Legs which are the next heaviest parts, the movement of which means the patient now forms an arc (Fig. 50).
4. The Trunk. This move is done either by inserting your hands under the waist and upper thighs and lifting the pelvis or by

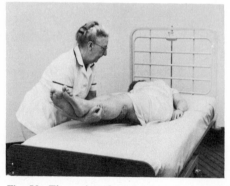

Fig. 50. The patient forms an arc when the head and legs have been moved to one side and he or she should try to help to move the pelvis.

asking the patient to dig the head, shoulders, hands and heels into the bed to 'bridge' the pelvis and lift it to the new position.

Turning a Patient Over

The manoeuvre of turning a patient over normally requires two people—one at each side of the bed. Tell the patient which way you are going to do the turn. Then warn the patient that he or she will be moved to the *opposite* side of the bed first. For example, a left turn from the supine position in the middle of the bed without a movement across the bed to the right will mean the patient ends up on the left edge of the bed. Repositioning on the bed is much easier if it is done with the patient lying on the back.

USE THIS ORDER OF MOVEMENT IF POSSIBLE

Lift and turn the head as this tells the patient where he or she is going. Put the arms across the trunk to make a 'tidy parcel'.

Cross the leg on the side nearest to you over that furthest from you as this helps to promote the turn. Then turn the trunk. If you are lifting alone start at the side of the bed

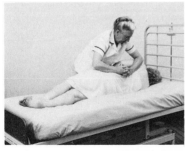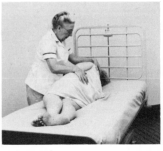

Fig. 51. *(left)* Turning the patient using the hands and forearms with the lifter kneeling on the bed.

Fig. 52. *(right)* Turning the patient using the relaxed hand grasp on the shoulder and hip.

Selection of the Lift

towards which you must first lift the patient towards yourself. Then

either
Turn the head, adjust the arms and cross the legs and move to the other side of the bed before turning the trunk

or
Move to the other side of the bed first and then proceed as above.

A long heavy trunk is best turned by placing your elbows just behind the patient's shoulder and pelvis and your hands at the waist (Fig. 51). If you are small you may have to kneel on the bed to do this. The advantage of kneeling on the bed is that the patient is then supported by your thighs after the turn has been done. A lighter patient may be turned by using the relaxed hand grasp on the shoulder and pelvis (Fig. 52). One forearm can then be slipped along the length of the trunk to hold it still until the legs are bent and the pillows placed for stability and support.

Selection of the Lift

PART 2 THE LIFTS

Shoulder (Australian) Lift

UP A HIGH BED (Figs 53–54)

The Shoulder Lift is the safest lift for two lifters to use to move a patient who can sit up and should be the lift of first choice in most situations.

Position the Equipment

Ensure the bed brakes are on. Turn back the bedclothes, remove obstacles behind the patient.

simultaneously

Inform and Position the Patient

The patient should be sat up in bed with straight or only slightly bent legs. Tell the patient that the lift up the bed will be done in small stages and ask him or her to keep the legs straight. If the arms are difficult to manage the patient may put them onto the lifters' backs but should not spread them too widely, or the lifters can use the arm position advocated for a single person lifter (see Fig. 60).

Position of the Lifters

Agree on a hand grasp with your helper. Both of you stand, one at each side of the bed, just behind the patient and facing the bed head. Put your outside feet forward one pace to point to the bed head. Turn your trunks and inside feet so that they point across the bed and at each other. Bend your hips and knees and keep your back straight. Put your inside shoulders under the patient's axillae from behind and your inside forearms under the patient's thighs by pressing your

Shoulder (Australian) Lift

hand into the mattress in order to get your hand under the thighs initially and then as near the buttocks as possible (Fig. 53). Grasp hands and put your outside hand on the bed level with your now forward feet. Your elbows should be bent (Fig. 54).

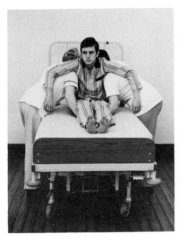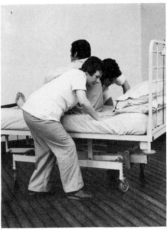

Fig. 53. *(left)* The shoulder lift up a high bed. Note all 3 heads are up and the lifters are braced for the lift. The patient's knees are nearly straight.

Fig. 54. *(right)* The shoulder lift up a high bed seen from the side. Note the bent elbows, hips and knees and the lifters' stance.

Commands   READY, BRACE, LIFT, and BRACE AND LIFT repeated as needed.

Execution   Both lifters straighten their knees, hips and forward elbows so that the patient is lifted towards their hands on the bed and lowered before reaching them. Both of you move your front feet forward then your rear feet and your hands on the bed an equivalent distance to that which the patient has been moved.

Special Point   A long lift (i.e., over a great distance up the bed) is much more hazardous to the lifters so do use short stages. The lift should only be up to the forward hand and foot of the smaller lifter if there is disparity of heights.

Shoulder (Australian) Lift

UP A LOW BED (Fig. 55)

The Shoulder Lift for moving up a patient in a low bed only varies from the preceding lift in the position of the lifters and the time between each phase of the lift.

Position of the Lifters

Agree on a hand grasp with your helper. You stand one at each side of the bed, behind the patient and facing the bed head. Put your outside feet forward one pace, to point to the bed head. Put your inside knees on the bed

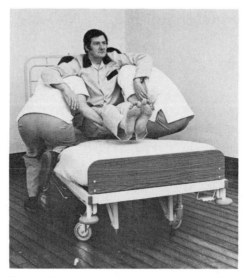

Fig. 55. The shoulder lift up a low bed. The lifters half kneel on the bed.

Shoulder (Australian) Lift

slightly to the rear of but beside the patient's buttocks and bend your knees fully so that you sit on your heels. Put your shoulders of the same side under the patient's axillae from behind and your hands under the thighs to grasp one another. Put your outside hands on the bed level with your outside feet (Fig. 55). Your knee and hip of this leg will now be bent to roughly right angles (90°).

NB A taller lifter may have the foot and hand further forward than a smaller lifter, *but* the distance the patient is lifted is to level only with the forward hand of the smaller lifter.

Execution   Execution is exactly as in the preceding lift and to the same commands, but your hips and knees only partially straighten. A little more time is taken between each phase of the lift as both lifters must move their rear hands, their feet on the floor, then their knees on the bed and rebrace ready to lift again.

Shoulder (Australian) Lift

UP A LOW DOUBLE BED (Fig. 56)

If possible, when using the shoulder lift up a low double bed, the patient should be moved near to one edge of the bed. The smaller or more agile of the two lifters kneels on the bed as this lifter must be able to sit back onto both heels. Both lifters proceed as in the previous lift but the lifter kneeling on the bed has both knees to the side of and behind the patient's buttocks and lifts mainly by slightly straightening the hips, knees and elbow (Fig. 56). If the bed is very soft this lifter or both of you should use hand blocks.

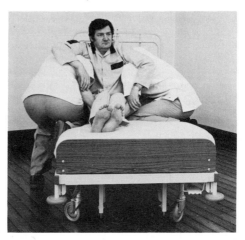

Fig. 56. The shoulder lift as performed up a low double bed. One lifter kneels completely on the bed.

Shoulder (Australian) Lift

Shoulder (Australian) Lift

FROM THE FLOOR TO A CHAIR
(Fig. 57)

Lifting a patient from the floor to a chair is quite difficult and should not be attempted by smaller lifters on a tall patient. The lift is made easier if it is done in stages on to seats and stools of increasing height.

Position the Equipment

The chair or stool to which the patient is to be lifted should be behind him or her when sitting up. If the patient is to be lifted first on to a lower stool then this is put behind the patient's back and the patient's own chair behind that (Fig. 57). Be very careful that you do not scrape the patient's back on the stool or chair as you perform the lift.

simultaneously

Inform and Position the Patient

The patient should be sitting up. The legs should be straight if the patient cannot help, but if he or she can help one or both knees should be bent so that the heels are able to dig into the floor. Tell the patient 'We are going to lift you (in stages)' and to push with the heel or heels when you say 'push'.

Position of the Lifters

Both lifters kneel or half kneel, that is, their nearside legs are in full kneeling and their outside legs are bent only so that their feet rest flat on the floor. The grasp is taken as in the Shoulder Lift up a low bed (see Fig. 55) for nearside hands. The lifters' rear/outer hands must rest on the front and outer corner of the stool (Fig. 57).

Fig. 57. The shoulder lift from floor to stool. Note the leg and hand position.

Commands	READY, BRACE, PUSH (to the patient) LIFT AND SWING.
Execution	Both lifters use a straightening action of their elbows and legs, rising as they do so. You must *not* let go of the stool or it may topple, catch the patient's back or slide backwards. On SWING you should swing the patient on to the front of the stool seat. The distance the patient travels backwards from floor to stool must be less than the distance to the lifters' outer feet.
Special Point	The lifters must totally reposition themselves to continue the lift to a higher level as they need no longer kneel. They should squat instead.

The Shoulder Lift as a Transit Lift

WHEELCHAIR TO CHAIR OR BED OR REVERSE (Figs 58–59)

The Shoulder Lift is also used to carry someone to another area. It is a safe and comfortable lift as the lifters have free hands with which to manage doors or other obstructions.

Position the Equipment

Ensure that the patient is free of bed clothing or rugs and is not anchored in any way to the present seat. The seat from which the lift is to be made must be stable and the seat on which the patient will be put down should be stable and have free access to both sides. This seat may not even be in the same room.

simultaneously

Inform and Position the Patient

The patient should be told where he or she is going and shown the new seat if it is in the same room. The patient should move to the front of the present seat so that half the length of the thighs is unsupported. Ask him or her to remain still and trust you.

Position of the Lifters

If the take off is from a low seat, both lifters should be either half-kneeling or half-squatting, obeying the principles of a wide base, and standing across the rear corners of the patient's seat. Insert your shoulders from behind under the axillae, grasp each other's hands under the patient's thighs and put your rear hands on the seat of the patient's chair or on the chair back if it is low and you are tall.

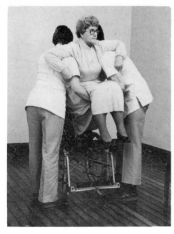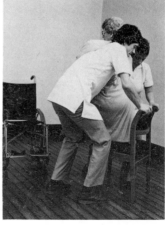

Fig. 58. *(left)* The shoulder lift from the wheelchair. The patient was sitting near the front of the chair.

Fig. 59. *(right)* The shoulder lift to a chair. The lifters' free arms hold the back of the chair and the elbows, hips and knees bend during the lowering phase.

Commands	For the lift: READY, BRACE, LIFT and STAND UP, WALK and/or TURN as needed.
Execution	The lifters brace and stand up by pushing on the chair and straightening their hips, knees and backs. When you are standing erect **NB** BOTH MUST STAND UP FULLY ERECT one of you may put one of the hands which pushed off from the seat on the patient's back as a stabilizing hand. The other lifter keeps a hand free ready to deal with doors or other obstructions (Figs 58 and 59).
Commands	For the lower: HOLD THE CHAIR BACK, READY, LOWER.

Execution   The lifters stand on each side of the new seat, outer feet forward, grasp the chair back and lower the patient, bending their hips, knees and ankles and the elbow of the fixed arm. The patient should be sat on the new seat by only half the length of the thighs. When you release your hand grasp on each other, one of you supports the patient from in front and tells him or her to rock back into the chair, or the other lifter goes behind the patient to perform a through arm grasp and lift to get him or her back in the chair.

Shoulder (Australian) Lift

FOR ONE LIFTER (Fig. 60)

The shoulder lift for one lifter can be used when the patient is capable of using one arm and possibly one leg to thrust on the bed, but still needs help on the other side, for example, a stroke-disabled patient.

Position the Equipment
Position the equipment as you would for the two-person lift but give the patient a hand block in the able hand.

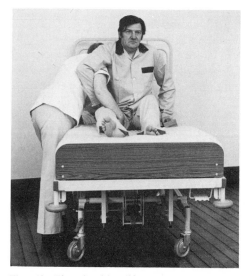

Fig. 60. The shoulder lift carried out by one person with the patient helping with sound arm using a hand block. Note the lifter has tucked the patient's forearm between the chest wall and her upper arm.

Shoulder (Australian) Lift

simultaneously

Inform and Position the Patient	Give the patient some instruction on thrusting on the hand block and make him or her aware that this will cause one buttock to lift off the bed. Then teach the patient to bend the sound knee and to thrust into the bed with that heel at the same time as he or she pushes on the hand block. The buttock on that side should rise even more.
Position of the Lifter	Take up the position ready to start the lift as previously described but put your arm which is to go under the thigh through between the patient's chest and upper arm and tuck the forearm between your chest and upper arm. Use the relaxed hand grasp under the near side thigh with your hand (Fig. 60) inserted under the thigh either from the outside or the inside.
Commands	READY, BRACE, PUSH, KNEE BEND, HAND BACK AND PUSH repeated as necessary.
Execution	The patient thrusts on the sound heel and hand so that the knee and elbow straighten, which helps the move up the bed.

simultaneously

You must straighten your legs and rear arm to lift the other side up the bed. The actions must be in unison or the patient will wobble. After the first move adjust your own feet and hand on the bed and ensure that the patient's foot and hand with the hand block have been moved back. Then both of you push again.

Special Point	The action can be so strong that 'overshoot' may occur and the patient should be warned to attempt only short jumps. Describe these as 'Little Kangaroo hops'.

Shoulder (Australian) Lift

Through Arm Lift

UP A CHAIR (LOW OR HIGH BACK) OR A WHEELCHAIR (Figs 61–63)

The Through Arm lift is most useful provided the patient's elbows can be bent to 90°. It may be used to lift a patient from a slumped position in a chair or wheelchair, up a bed, off the floor or from the bath. For the last two lifts two people are needed.

Position the Equipment

The wheelchair brakes should be on and any type of chair must be accessible from behind.

simultaneously

Inform and Position the Patient

Check first that the knees are bent if possible and, if the patient is in a wheelchair, that the feet are on the footrests. Assist the patient into the upright position by pushing on the back of the head or both shoulders with the relaxed hand grasp (see Fig. 23.) Tell the patient 'I'm going to lift you up in your chair.' 'Help me to lift your feet back' (onto the footrests). 'I'm going to lean you forward and then sit you up.'

Position of the Lifter

For a Lift in a Wheelchair or Low-backed Chair (Fig. 61)

Stand behind the chair, put your feet into a walk standing position with the knee of your forward leg resting firmly against the back of the chair. Your front thigh may also be in contact with the chair. Your knees and hips should be bent and your weight on your forward foot.

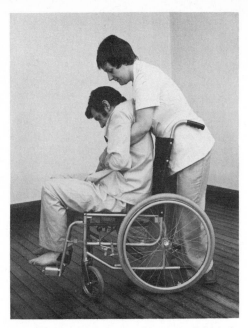

Fig. 61. The through arm lift up a wheelchair. Patient's feet on the footrests; lifter supporting her forward thigh on the wheelchair back.

For a Lift Up a High-backed Chair

A lift up a high-backed chair is feasible for a tall lifter positioned as in the lift for the wheelchair or low-backed chair but a shorter lifter must use a safe footstool. The placing of the stool should be with the maximum length available to the feet in the walk standing position as in Figure 63, where a square stool is placed obliquely to the chair back. Initially and in order to sit the patient up

either
Put both hands under the back of the patient's head, thumbs adjacent, to cup the occiput and push the patient upright (Fig. 24)

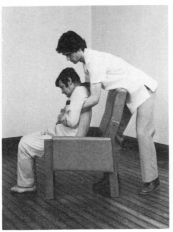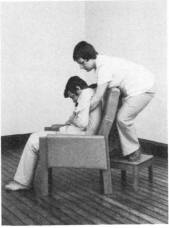

Fig. 62. *(left)* The through arm lift up a high-backed arm chair. Note that a tall lifter can reach and does not have to lean on the chair back.

Fig. 63. *(right)* The through arm lift up a high-backed armchair done by a shorter lifter who must use a footstool. Note the position of the stool to give maximum foot space.

or

Use the relaxed hand grasp on the back of both shoulders.

Next slide your hands between the upper arms and chest wall, turn your palms downwards and grasp the patient's forearms. Keep your own elbows straight. If the patient can grasp the wrist of one hand with the other hand you should grasp his or her forearms near the wrists (see Fig. 25). If the patient is unable to help in this way grasp as near to the patient's elbows as possible and fold the forearms across the chest. (see Fig. 26). Put your head to one side over the patient's shoulder and keep your head up.

| **Commands** | To the patient: GRASP ONE WRIST WITH THE OTHER HAND. |
| | To yourself: READY, BRACE, LIFT. |

It is important to brace the 'slack' out of both your own and the patient's positions before attempting the actual lift.

| **Execution** | After bracing yourself, lift by straightening your hips most and your knees less. |

| **Special Point** | The patient's buttocks should be lifted enough to clear the seat as there will be an automatic swing back onto the seat as the lift is executed. |

Through Arm Lift

Through Arm Lift
(One Lifter)

UP A BED (Figs 64–65)

A lifter who is alone with a fairly helpless patient who has slipped down the bed should use the Through Arm Lift. It looks startling to those who are unaccustomed to working in community or district roles, but it is much used in the home.

Position the Equipment

Put the bed brakes on and remove the pillows from behind the patient who should be sat forward as you do so.

simultaneously

Inform and Position the Patient

Sit the patient forward using one of the lifts shown on pages 127–9, and ask him or her to grasp one wrist with the other hand if

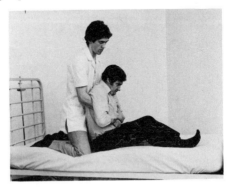

Fig. 64. The through arm lift up a bed done by a taller lifter in kneeling.

Through Arm Lift

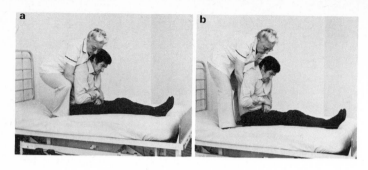

Fig. 65. The through arm lift up a bed done by a smaller lifter in standing. (a) Taking up the stance in squat standing and ready to lift and (b) the lift completed and the feet moved back before squatting again.

possible and also to bend the knees ready to push with the heels if this is feasible.

Position of the Lifter

Remove your shoes. If you are taller than the patient kneel on the bed behind him or her and a little distance from the buttocks. (Fig. 64). Put your head to the side so that it is over one of the patient's shoulders.

If you are small stand on the bed behind the patient with your feet apart more than the width of the patient's buttocks, and pointing slightly outwards. Bend your hips and knees sufficiently to take the through arm grasp (Figs 65a and b).

Commands

READY, BRACE, PUSH WITH YOUR HEELS (to the patient).

Execution

Bracing of slack out of both your own and the patient's body is essential before attempting the lift, in which you straighten your hips if

Through Arm Lift

you are kneeling, and your hips and knees if you are standing.

If you are kneeling, the distance of the lift is controlled by your short distance away from the patient. In standing, a lift over a greater distance is feasible because you can swing the patient back between your feet, but this is likely to cause you to fall over forwards on top of the patient. Short distance lifts are advisable.

Through Arm Lift

FROM FLOOR TO BED (Fig. 66)

The through arm lift from floor to bed requires two people and is used for a patient who is unable to help by using his or her legs. The tallest lifter must lift the head and trunk.

Position the Equipment

The bed brakes must be on and bed clothes removed. The bed should be no more than two paces away from the patient and parallel with him or her.

simultaneously

Inform and Position the Patient

Sit the patient up with the legs straight out in front. Ask him or her to grasp the wrist of one arm with the other hand if possible. Tell the patient 'We are going to lift you and put you on the bed. We shall stand up first then swing you slightly.'

Position of the Lifters

The lifter behind the patient gets into half-kneeling or squatting. This lifter pushes the patient's head forward a little (see Fig. 25) and takes the through arm grasp (see Fig. 26) with the head to one side and over one of the patient's shoulders. The other lifter kneels or squats with one foot in front of the other, beside the patient's knees, puts one hand under the thighs and one hand under the lower calves (Fig. 66a) in the relaxed palm grasp.

Through Arm Lift

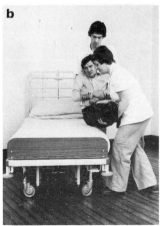

Fig. 66. The through arm lift from the floor to bed.
(a) The starting position for each lifter is different.
(b) Stand, swing, walk and lower.

Commands   BRACE, LIFT AND STAND, SWING, WALK AND
DOWN.

Execution   Both of you must stand up together; and
immediately the lifter of the legs steps
forward while the lifter at the head walks
sideways to the bed, at the same time
imparting a swing to the movement so
that the patient is no longer a dead weight
(Fig. 66b). The swing must be high enough to
allow the patient to clear the edge of the bed
and be placed on it. Adjustment of position to
the middle of the bed can then take place
using another lift.

Through Arm Lift
(Two Lifters)

INTO AND OUT OF THE BATH
(Figs 67–69)

Lifting a very helpless patient into and out of the bath is a manoeuvre which should be done using a mechanical hoist if possible. If it must be done by a manual lift then the Through Arm Lift is the best choice, and two or three lifters will be needed. Putting the patient into the bath is less stressful than taking him or her out of the bath. The tallest lifter takes the head and shoulders and must be taller than the patient, the second lifter takes the legs.

A third lifter, if needed, takes the thighs and supports them as near to the buttocks as possible. This lifter would then have to half-kneel on the stool as the patient is lifted off it onto the edge of the bath and into the bath, being careful not to tip the patient towards the bath.

Position the Equipment

Fill the bath to a quarter or a third of its depth with water which will be tolerated by the lower half of the body. (Cooler than hand-hot.) Use a nonslip mat in the bath. Remove soap trays and other obstructions. Place folded towels along the free edge of the bath. Place a stool the height of the bath edge alongside the bath at the level at which you wish the patient's bottom to go in the bath and place a folded towel on it.

Fig. 67. *(left)* The through arm lift into the bath. Note the lifters' positions as they are ready to lift.

Fig. 68. *(centre)* Momentary rest on the bath side to let the trunk lifter transfer his or her own weight into the bath.

Fig. 69. *(right)* The lowering. The leg lifter must squat.

simultaneously

Inform and Position the Patient

Undress the patient. Use the through arm lift, with a helper to support and lift the legs and take the patient from the wheelchair on to the stool. Ask the patient to lean to the bath and feel the water temperature with the nearside hand, and then to grasp one wrist with the other hand if possible.

Position of the Lifters

(The lift is described for two lifters and for a bath with access only at one side.)

The lifter of the legs must have bare forearms and be prepared to lift both straight legs at once to just above the level of the bath edge and then to lower them into the water. This lifter must be prepared to take a step forward and then perhaps kneel or squat beside the bath as the legs are lowered into the water (Fig. 67).

Through Arm Lift

The lifter of the trunk stands behind the patient, with the nearside foot in the bath (first remove stockings, socks etc.). Ensure that your foot is on the nonslip mat. Your other foot should be behind the middle of the patient and pointing towards the front and outside of the stool. Take the through arm grasp.

Commands READY, BRACE, LIFT, SWING, LOWER, LIFT, SWING, LOWER.

Execution If the patient is very light the lift may be done in one swing and the lifters raise the patient to a height sufficient to clear the bath side then lower him or her into the water.

If the patient is heavy the lift should be done in two stages. The lifter of the trunk gives the commands and you raise the patient on to the bath side, where you perch him or her momentarily before lifting again and lowering the patient into the water (Figs 68 and 69). The lifter of the legs must synchronize the movement of the patient's legs over the bath edge and must not lower the weight of the legs on to the bath edge. The pause is to allow the lifter of the trunk to transfer his or her own weight fully from the foot outside the bath to that inside the bath.

Taking a patient out of the bath is a reversal of the above procedure. Dry as much of the patient as you can and let the water out first if you don't want to get very wet during the lift. On the other hand the buoyancy of the water imparted to the body can assist in initiating the lift and may be more helpful in lifting a very heavy patient.

Special Point If there is access to the head of the bath it is unnecessary for one lifter to put a foot in the bath as the lifter of the trunk can stand behind the head of the bath. The patient can be allowed to slide down the slope once the buttocks have reached the water.

Orthodox Lift

UP A BED (Figs 70–71)

The orthodox lift is often used when the shoulder lift would be better. This lift should only be used when a patient may not have pressure on the axilla or chest wall, as following mastectomy or when totally helpless. It is a lift which causes high truncal pressures on the lifters. Two lifters are always needed to lift an adult, and more than two are needed if the patient is heavy.

Position the Equipment

Check that the bed brakes are on. Turn back the bedclothes completely to leave the patient free to be moved. Remove or rearrange potentially obstructing pillows. Ensure that the mattress is level with your mid-thighs.

simultaneously

Inform and Position the Patient

Ensure that the patient is in the middle of the bed. If not, and he or she cannot be moved to the centre then the tallest lifter should be on the side which requires the longest reach; that is, the smallest lifter stands on the side of the bed to which the patient is nearest. Tell the patient what you intend to do. 'We are going to lift you up the bed, but first we want you to cross your arms in front of your chest and we shall put our arms under you. Can you lift your head?' 'Good, now when I say HEAD UP I want you to raise your head' (Fig. 70). Alternatively, the patient may sit up to be lifted (Fig. 71).

Orthodox Lift

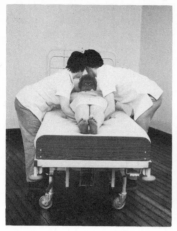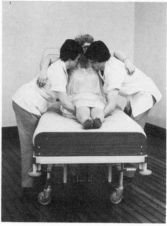

Fig. 70. *(left)* The orthodox lift up a bed for a patient in lying position who can raise the head.

Fig. 71. *(right)* The orthodox lift up a bed for a patient who can sit up.

Position of the Lifters

The lifters stand at each side of the bed in the lunge stance with their forward feet nearest the bed head pointing up the bed. Both of you bend your hips and knees. Each lifter inserts the arm which is nearer the bed foot under the patient's thighs; each grasps the other by a previously agreed grasp, and both wriggle their arms as high under the thighs as possible. Then each of you inserts your other (forward) arm under the patient's waist and tries to use a double grasp on each other's forearms avoiding a hand grasp if possible.

Commands

To everyone: READY, BRACE. To the patient: HEAD UP. To the other lifter: LIFT AND LOWER.

Orthodox Lift

Execution Immediately after the patient's head is raised, you raise the patient by both of you straightening your hips and knees and leaning your trunks slightly backwards and the patient is swung up but NO FURTHER THAN YOUR FORWARD FEET. The patient is lowered and if further movement is necessary your feet are repositioned, your hands remain in position and further short lifts are performed until the patient has been moved as far as is necessary.

Special Points If the patient cannot raise the head a third helper may raise it. When lifting small patients your forward arms may go under the head and cross on the upper back with your other hand under the thighs.

Orthodox Lift

BED TO CHAIR OR REVERSE
(Fig. 72)

Like the orthodox lift up a bed the lift from bed to chair or reverse should only be used when less stressful lifts are not feasible. Two lifters are needed.

Position the Equipment

Put the bed brakes on. Remove obstructing bedding and then position the chair as near as possible without hampering your movements. Fix the chair against something rigid.

simultaneously

Inform and Position the Patient

Prepare the patient in suitable clothing, for example, help to put on footwear and a dressing gown if he or she is to sit out of bed. Turn the patient to sit on the side of the bed so that half the length of the thighs is unsupported. Tell the patient where he or she is going and to put his or her arms across your shoulders.

Position of the Lifters

Stand one on each side of the patient in the lunge position with your forward feet pointing under the bed and your other feet pointing towards each other. Position your nearside arms by taking the forearm or elbow grasp behind the patient's waist (Fig. 72). Then bend your trunks sideways and down away from the bed, at the same time bending your hips and knees and grasp under the patient's thighs with your outer hands. Brace upwards and turn your bodies more to face each other, pivoting your feet also so that

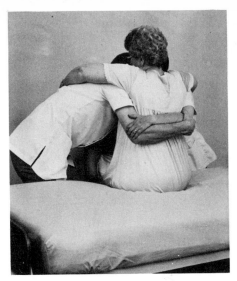

Fig. 72. The orthodox lift off a bed to a chair. Note the position of patient's thighs which are half off the bed and the lifters' comfortable grasp on each others' elbows.

your previously rear feet now point to each other and the previously forward feet point away from the bed. This initial pivot is important as it realigns the lifter's bodies in preparation for the load

Commands READY, BRACE, LIFT, WALK AND LOWER.

Execution On LIFT both lifters stand up *fully erect* and lift the patient clear of the bed. On WALK you proceed to the new seat and stand in front of it so that the patient's buttocks are over the front edge of the chair. On LOWER pivot your weight from both feet on to that nearest to the chair and at the same time bend your hips and knees until the patient rests on the chair.

DO NOT TRY TO PUT THE PATIENT TOO FAR INTO THE CHAIR.

Both let go but one lifter should move to stand in front of the patient.

Either

One lifter holds the patient and the other goes round the back of the chair to use the Through Arm Lift to take the patient more securely into the chair

or

The patient is advised by the lifter in front and helped to get further back into the chair by hitching alternate hips backwards or by rocking.

In performing this lift from chair to bed, first get the patient to the front of the chair by methods advocated in Chapter 4 so that half the thighs are unsupported by the chair. Block the patient's feet by one each of your feet and use the middle lifter's stance for the three-person lift from a chair (see p. 96). BRACE, STAND UP, WALK AND TURN. Put the patient on the bed, so that half the thighs are unsupported by the bed and immediately, one of you blocks the patient's position by moving in front and leaning your body forwards to prevent the patient from falling forwards.

The further move on to the bed may be accomplished

either

By the patient using hand blocks

or

By your using the shoulder lift

or

By one of you using the through arm lift from behind the patient by kneeling or standing on the bed

or

By turning the patient to put the legs on the bed and lifting or rolling him or her to a better position.

Raising a Patient's Pelvis for Nursing Purposes

TO INSERT A BEDPAN OR INCONTINENCE SHEET (Fig. 70)

At least two lifters are required. As this is a static lift it is very stressful on the back and should *not* be attempted by one person alone. If possible the patient should always be either rolled from side to side for these procedures or lifted out of bed as on to a commode.

Position the Equipment
Put the bed brakes on. Have the equipment to be used very close at hand.

simultaneously

Inform and Position the Patient
If the patient can help at all he or she should have the knees bent so that, if possible, the feet rest flat on the bed and/or the arms should be straight and on the bed beside the chest. Some patients may prefer to hold on to the head of the bed or an overhead handle.

Position the Lifters
Both lifters stand opposite the pelvis, feet either astride or one foot in front of the other. The remainder of their position is as for the orthodox lift (see Fig. 70).

Commands
If the patient cannot help: READY, BRACE, LIFT or If the patient can help: READY, BRACE, PUSH YOUR HEELS IN THE BED, LIFT.

Execution
Both lifters lean backwards transferring their weight from their forward to their rear legs. If a third person is available the nursing procedure is carried out as quickly as possible by this person.

If a third person is not present the stronger of the two continues to hold the pelvis up by bending your elbows and so tipping the patient nearer to yourself. The other lifter carries out the nursing procedure. The lifter holding the patient up must not be tempted to help by removing one supporting arm to assist, for example, straightening that side of the incontinence sheet. To lower the patient both lifters should support him or her and lower so that the full weight does not fall on one lifter.

Raising a Patient's Pelvis

Three-person (Team) Lift

STRETCHER TROLLEY TO BED AND REVERSE (Fig. 73)

A team lift usually involves three or four lifters. If three lifters are involved; one lifts the legs, one (the strongest) the pelvis and one the shoulders and head. This is the number of lifters used when the patient *does not* have life support attachments to the head and/or neck. If there is such equipment, and it must be carefully lifted and watched, then a fourth lifter is needed to look after the head and equipment and the the shoulders are lifted by one lifter, the pelvis by a second and the legs by a third lifter. In this case the lifter at the head stands behind the patient's head and at right angles to the other lifters if possible. This lifter is then likely to give the commands. For a three-person lift the commands should be given by one of the two lifters of the trunk or the lifter of the pelvis, though traditionally the lifter of the head and shoulders is more likely to give the commands. This lift is used to move a helpless patient on a bed, to turn a patient or to effect a transfer to or from a stretcher trolley.

Position the Equipment

The bed and stretcher trolley should be so positioned that the team can take two or three steps back, then pivot slightly as a group and walk forward to the new support. Limitations of space sometimes make side stepping necessary and, if space is confined, it may be necessary to put the bed and stretcher trolley side by side. It is then wiser to get a fourth person to pull the stretcher trolley out after the patient is lifted so that the team can then step forward to the bed.

simultaneously

Inform and Position the Patient

Even an unconscious patient should be advised about what is to happen as he or she may be less deeply unconscious than you realize and startling actions may cause unwanted movements. So tell the patient what you are going to do using the type of 'patter' advocated on page 52. If the patient is in the middle of the bed it is strongly advisable to move him or her first to the edge, after folding the arms across the trunk.

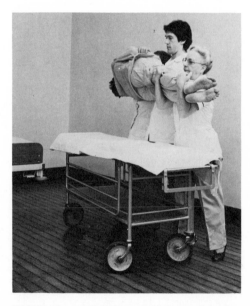

Fig. 73. Team (three-person) lift from a stretcher. Note the lifters' bent elbows which have tipped the patient to be over the base of the lifters' feet.

Position of the Lifters	All the lifters stand in walk standing with the same foot forward, hips and knees bent most on the forward leg and the rear leg straighter to maintain your balance. The lifter of the legs puts one forearm under both the patient's knees and one under the ankles. The lifter of the pelvis puts one arm as high as possible under the thighs sliding the arm up from under the knees and one as low as possible above the pelvis, sliding the arm first under the waist. The lifter of the head and shoulders puts one arm under the waist and slides it up under the thorax and the other arm between the top of the shoulder and the head so that the head is cradled on that arm and shoulder and the hand lies obliquely across the upper back.
Commands	READY, BRACE, LIFT, ELBOWS BEND, STEP BACK, TURN RIGHT (OR LEFT), FORWARD, LOWER.
Execution	The lifters raise the patient by straightening their legs and standing *erect*. They then bend their elbows so that the patient is held slightly tipped towards them (Fig. 73) and safely resting against their upper arms and trunks. In this position the patient's weight is transmitted to their feet. They step back, pivot or turn as previously agreed, step forward to the new support, put one foot in front of the other (the same foot for each of them) and, bending their hips and knees of the forward leg, lower the patient.

NB In Figure 73 the middle lifter was much taller than those on each side yet the patient was held straight.

Three-person (Team) Lift

BED TO CHAIR OR CHAIR TO BED
(Fig. 74)
To prepare for a lift from bed to chair follow
the procedure in the preceding lift until you
are alongside the chair with the patient
supported over it.

Then the middle lifter lowers first,
followed immediately by the leg lifter. When
the patient is in a sitting posture all three
lifters lower the patient to the seat. The leg
lifter must squat or kneel and all three must
be careful to bend their hips and knees. They
must also bend their backs slightly sideways
towards the patient's feet. They must **not**
deliberately rotate their backs or the patient
will corkscrew.

The lift from the chair starts differently but
finishes like the stretcher trolley to bed lift.

Inform and Position the Patient

If possible move the patient to the front of
the chair especially as this lift is most often
used to lift a patient from an armchair back
to bed.

Position of the Lifters

The position of the lifters' hands is the same
as in the preceding lift. The leg positions are
different but you must agree on which foot
will be forward. The lifter of the legs squats or
kneels with one foot slightly in front of the
other. The lifter of the pelvis half squats with
one knee and hip more bent than the other.
This lifter must also bend slightly sideways
towards the patient's feet. The lifter of the
upper trunk and head has most bend in the
leg nearest the patient's waist. The arm
positions are as for the Bed to Stretcher
Trolley Lift.

Three-person Lift

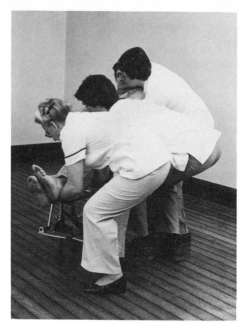

Fig. 74. Team (three-person) lift from a wheelchair. Note the tallest lifter in the middle has squatted to accommodate his longer legs.

Commands	READY, BRACE, LIFT AND STAND, ELBOWS BEND, STEP BACK, PIVOT, WALK, LOWER.
Execution	The middle lifter must initiate the lift assisted by the adjacent arm and side of the upper trunk lifter then, when the patient is gradually being straightened, all three lift together, stand up and tip the patient towards themselves by bending their elbows, step back, turn and walk to the new support, put one foot in front of the other and lower the patient.

Axillary Lift

CHAIR TO STANDING (Figs 75–78)

The axillary lift is a rocking lift in which repeated attempts are made to stand up until the standing position is achieved.

Position the Equipment

The chair from which the attempt is made should be firmly fixed

either
By putting the brakes on and lifting the footrests (wheelchair)

or
By putting the chairback against something solid, like a wall.

simultaneously

Inform and Position the Patient

The patient should be near the front edge of the chair, feet flat on the floor and under the knees which are bent to a right angle or more. Tell the patient to lean the trunk forward from the hips with a straight back so that the head is in line above the knees (Fig. 75). The patient's best hand is put on the thigh of the same side ready to push. The lifter should stand on the patient's most disabled side.

Now teach the patient what to do and how to help. Tell the patient to push on the good foot in order to 'feel' the floor, then to push with the good hand on the thigh in order to feel the buttock start to lift on that side.

When this has been done the patient should be told that the rise will be repeated and get bigger, with a sit down but no rest between each attempt (if necessary demonstrate by sitting in another chair yourself and performing the same actions). Reassure the patient that you will help and warn him or her to let go of the thigh with the hand as he or she stands up further.

Position of the Lifter

Put your forward foot across the front of the patient's toes and the inside of your knee of the same leg in front of the nearest or both of the patient's knees. Bend your own knee to do this. Put your rear foot at right angles to your forward foot alongside the chair in lunge (see basic foot positions p. 20). Take an axillary grasp (see p. 27 and also Fig. 27a and b).

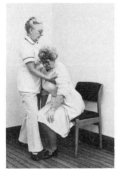

Fig. 75. *(left)* The axillary lift from sitting to standing. Note the blocking foot and knee, yet the lifter is to the side of the patient.

Fig. 76. *(centre)* The axillary lift almost complete.

Fig. 77. *(right)* The axillary lift. Note that the lifter's stance is very upright.

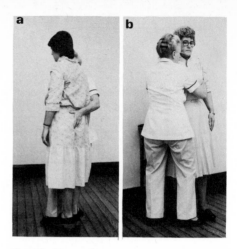

Fig. 78. The axillary lift accomplished. (a) The lifter is still blocking the foot and knee and is now pushing forward on the sacrum and backwards across the upper body. (b) The lifter's stance is still upright.

Your body weight should be over your rear foot initially so that your trunk is *beside* the patient and **not** in front him or her where you will prevent both any lift and the eventual stand up.

Commands

READY, PUSH, repeated every time the patient has just sat down.

Execution

The patient pushes down with the foot and hand and initiates the rise. The lifter exerts a gentle pull in the line of the back (i.e., forwards and upwards) and in exact time with the patient's action (Fig. 76). Do not transfer your own body weight forwards and back between your two feet as the patient moves forwards and up and sits again (Fig. 77). Each effort is repeated as many times as necessary and each should cause a greater elevation of the patient's buttocks from the seat. Eventually the patient should let go of the thigh with the pushing hand and as he or she becomes nearly upright you swiftly remove your rear hand from under the axilla, place it very firmly on the sacrum and push the patient hard forwards. At the same time brace the patient's shoulders backwards with your other arm and the knees with your supporting knee, thus giving three contra pressures which complete the process of standing (Fig. 78a). Your weight should now be mainly on your rear foot (Fig. 78b).

Special Points Do not be tempted to move your feet.

Discomfort under the patient's axilla indicates one of four errors:

1. The assisting pull is not in the line of the back.
2. The pull is being exerted before the patient initiates the push.
3. The lifter's hands have turned so that the radial edge of the fingers is being used for the lift instead of the whole palmar surface of the hand.
4. The tips of the fingers are digging into the axilla instead of being kept flat.

Axillary Lift

SITTING ON LOW BED TO CHAIR OR REVERSE (Figs 76–79)

The axillary lift for sitting on a low bed to a chair or reverse involves a similar procedure to the preceding lift but when the patient's buttocks rise sufficiently clear of the present seat rotating pressures are exerted on the axillae and a turn is executed so that he or she rises from one seat and sits down on the adjacent seat.

Position the Equipment

The two seats must be at right angles to one another, touching at one corner if possible and both fixed.

simultaneously

Inform and Position the Patient

As in the preceding lift, except that the patient is told where he or she will sit down and that *you will carry out the turning action.*

Position of the Lifter

As in the preceding lift.

Commands

As in the preceding lift.

Execution

Proceed as in the preceding lift and when the patient's buttocks are high enough to clear the present seat and any obstructions, at the peak of the 'rise' turn the farside axilla forwards towards you and turn the nearside axilla backwards away from you so that the shoulders and trunk turn. This is a very swift action and correct timing is essential. The pull on the axilla is gentle and swift and, if done in correct timing, will not cause discomfort.

Special Points   Keep your own trunk erect and turn the patient with your arms (see Figs 76 to 79). If the patient starts to help the turn by taking the buttocks round for himself or herself instruct him or her not to do so but to concentrate only on the up and down movement.

Axillary Lift plus Single-lifter Shoulder Lift

CHAIR TO HIGH BED (Figs 79–81)

A succession of manoeuvres is useful when a bed cannot be lowered. The patient will need to know how to use a hand block in the good hand and should have been made familiar with forward rocking and the axillary lift and turn previously described.

It is most important that the lifter stays very close to the patient, leaning against him or her to give support at the edge of the bed. There are three foot positions used: that for the axillary lift, that is, wide lunge, ten-to-two, and another wide lunge with the direction of the feet reversed. A succession of three grasps is also used: axillary, relaxed hand, and single-person shoulder grasp and lift.

Position the Equipment

Ensure that the bed brakes are on and turn the casters up and down the bed. Put the hand block on the bed where the patient can easily reach it as you lean him or her against the bed. Put the patient's chair so that one side touches the bed. Ensure that the chair is blocked from moving backards. If a stool or chair is available put it adjacent to the bed and opposite the patient's present seat for the patient to use as you tell him or her to put the foot on the stool and to push.

simultaneously

Inform and Position the Patient

The patient's best hand should be nearest the bed and he or she should sit near the front of the chair and lean forward (see axillary lift).

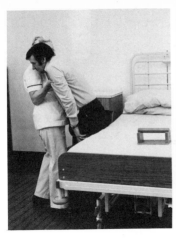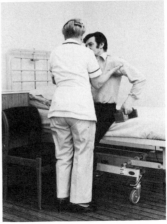

Fig. 79. *(left)* The axillary lift and turn to a bed. If the bed is low the patient just sits down.

Fig. 80. *(right)* The axillary lift completed onto a high bed. The lifter leans onto the patient to block the legs.

Warn the patient to grab the hand block as you lean his or her buttocks on the bed edge. Tell the patient that you will help him or her to rock up, that you will execute the turn and protect him or her from a fall at all times.

Position of the Lifter

Initially the lifter is positioned for the axillary lift. As the patient is leaned on the bed, you take the ten-to-two stance to stop his or her feet slipping forwards and your knees and hips should be straight and leaning against the patient's legs. Try not to lean hard on the bed.

Commands

PUSH, HOLD THE BLOCKS, FOOT ON THE STOOL, PUSH

The commands are successively those for the axillary lift followed by the shoulder lift.

Axillary Lift

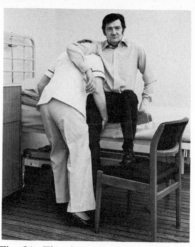

Fig. 81. The single-person shoulder lift and the patient using a block and a chair are used to lift the patient safely onto a high bed.

Execution

The axillary lift is performed until the patient has gained sufficient height for you to turn the buttocks to rest against the bed (Fig. 79). Then swiftly move your most distant foot beside your other foot into the ten-to-two stance. At the same time lean your legs against the patient's legs. Help the patient to get the hand block and to lean on it with the good hand. Pull the spare chair or stool beside you so that the patient's foot on that side can be lifted on to it (Fig. 80). Help the patient to do this using the relaxed hand grasp with your hand on the same side as the patient's leg (i.e., your right hand on the patient's left leg). Now put the same hand under the other thigh using the relaxed hand grasp from inside the patient's thigh. Lean away from that hand and put that shoulder under the axilla of the weakest arm taking up the single-person shoulder grasp. Tell the patient to push on the hand block and the foot and assist him or her further on to the bed (Fig. 81).

Axillary Lift

HIGH BED TO CHAIR (Figs 79–80)

The axillary lift from a high bed to a chair is a succession of manoeuvres in which the patient is first brought to the edge of the bed, helped to stand momentarily, then quickly turned to sit in the chair.

Position the Equipment
Put the bed brakes on and fix the chair so that the seat is at right angles to and touching the side of the bed.

simultaneously

Inform and Position the Patient
Ask the patient to sit up and assist him or her to turn to sit on the side of the bed. Ensure that he or she is wearing shoes that fit.

Position of Lifter
Stand in front of the patient with your feet in the ten-to-two stance and your upper thighs leaning against the patient's knees. Put your hand on the side away from the chair into the back of the appropriate axilla, and your other hand into the front of the patient's other axilla (Figs 79 and 80).

Commands
REACH DOWN FOR THE GROUND WITH YOUR BEST LEG, STAND ON IT, NOW REACH WITH THE OTHER LEG AND STAND ON IT.

Execution
Tell the patient to slide his or her best foot in front of your legs and reach for the ground. Then tell him or her to reach down with the other foot. The feet should be on the ground between your feet. Next rapidly move your

own pelvis backwards away from the patient
and at the same time apply rotating pressures
on the patient's axillae as in the preceding
lift. Take your own weight on to your leg
furthest from the seat and the patient will sit
down.

Elbow Lift

WHEELCHAIR TO TOILET SEAT OR REVERSE (Figs 82–84)

The elbow lift is for use on a very incapacitated patient who can sit in a wheelchair but cannot help in order to get out of it. It was devised for patients who must use an onward facing toilet in limited space and can be performed on very heavy patients who have some muscle tone. This lift is unsuitable for very flaccid (floppy) patients or those whose hips must not be bent more than 90°.

Position the Equipment

The chair from which the patient is to be moved should be positioned with the front adjacent to the seat on to which the patient is to be moved and the distance between the two seats is critical. Measure the length of the patient's thighs from the front of the knee to the back of the buttock. Then measure from the back of the knees to the middle of the new seat. If these two are the same, the distance is correct, as the patient will be pivoted on his or her own feet in the course of the lift. If possible remove the chair arms, but if this is not possible a lift over the chair arm is still feasible. Lift the wheelchair foot supports. Ensure that the brakes are on.

simultaneously

Inform and Position the Patient

Warn the patient of what is to happen then use a lateral rock to bring the patient to the front edge of the chair. Put the feet on the floor side by side and inside the wheelchair footrests. Take down the clothing if the patient is to be put on the toilet, as trousers or pants help to hold the thighs together. Tell the patient you will use a forwards and backwards rock once or twice until the buttocks are high enough to allow a turn to the new seat.

Position of the Lifter

Stand at the corner of the chair with your forward foot across the patient's feet and the knee of the same leg across his or her knees. Put your other foot outside the front wheel of the wheelchair. Turn your body to face the patient. Lean the patient forward and support the nearside shoulder on your lower trunk. With your nearside arm reach across the back of the neck, in front of the opposite shoulder and grasp under the farside elbow. Grasp under the nearside elbow with your other hand. Tuck the elbows well in to the waist more towards the back (Fig. 28a and b).

Instruct the Patient

Rewarn the patient of what will happen and ask him or her not to help you.

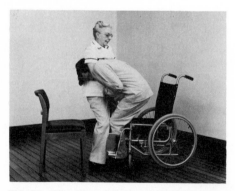

Fig. 82. The elbow lift from wheelchair to chair through 180 ° to show positions of both patient and lifter. Note the leg positions.

Commands   Brace yourself and take up your own and the patient's slack.

Execution   Rock the patient forwards by transferring your weight on to your rear leg (i.e., the leg in front of the feet and knees) and continue to rock back and forwards, building up the size of the lift at each rock until you have gained enough height for the patient's buttocks to clear the seat—and chair arm (Fig. 82). At the height of your final rock swing the patient round by turning your body towards your *farside* arm and moving your weight on to your *nearside* leg (Fig. 83). Lower the patient at the speed of the previous rocks, but on to the new seat letting go of the most distant arm as he or she sits (Fig. 84). Let go of the other

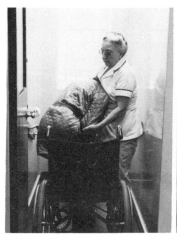 

Fig. 83. *(left)* The elbow lift from wheelchair to toilet seat, half way round.

Fig. 84. *(right)* The elbow lift nearly completed The lifter lets go with the farside hand as the patient is sat down.

arm and transfer your hands to the shoulders to stabilize the trunk. Be prepared to adjust the clothing before the toilet is used.

Special Points   The patient's feet may have crossed at the forefoot or ankles. Uncross them after turning the patient or cross the feet before turning the patient. The foot on the side towards which the buttocks will turn should be crossed in front of the other one.

Elbow Lift

LOW BED TO CHAIR OR REVERSE
(Figs 28 and 82)

The elbow lift may be performed through 90°
if it is possible to put the chair alongside a
low bed or other seat.

Position the Equipment — Put the brakes on both the chair and the bed
and then remove the wheelchair arms. The
chair should touch the side of the bed. Raise
the footrests on the chair.

simultaneously

Inform and Position the Patient — Tell the patient that he or she is to be moved
to the chair and that you will lift him or her
by rocking until there is enough height to
allow the turn. Use lateral rocking to get the
patient to the edge of the bed and put the feet
side by side inside the area of the wheelchair
footrests.

Position of the Lifter — Stand at the opposite side from the
wheelchair and put your outside foot across
the front of the patient's feet and the knee of
the same leg across the knees. Your other foot
should point across the bed. Use the elbow
grasp (see Fig. 28a and b).

Commands — To the patient: DO NOT HELP.

Execution — Lean the patient forward to rest the nearside
shoulder against you. Take the elbow grasp,
brace and transfer weight from your foot near

the bed to your foot inside the chair footrests. Rock increasingly until the patient's buttocks rise free of the bed and high enough to clear the wheel of the chair (see Fig. 82). Turn the patient by pulling the farside elbow nearer to you and the other one slightly away from you. Pivot your own hips as you do so and sit the patient down on the wheelchair. Go behind the patient and use the through arm grasp to lift him or her further back into the wheelchair.

Elbow Lift

Waistband Lift

CHAIR TO STANDING (Fig. 85)

Patients who are wearing clothes which have a fitted waistband or a firm belt (Fig. 84) may be lifted from sitting to standing using this lift.

Position the Equipment

Fix the seat firmly or put the brakes on.

simultaneously

Inform and Position the Patient

Ensure that the patient is wearing shoes or firm slippers. Tell the patient that he or she is going to stand up. Move him or her by using lateral rocking to the front edge of the chair, feet together and ready to lean forwards. The arms should

either

Rest over your shoulders (see Fig. 86)

or

Be supported by grasping your waist or tucking the thumbs into your waistband.

Position of the Lifter

Stand in front of the patient with your feet at ten-to-two on each side of his or her feet and your knees bent outward on each side of the patient's knees. Lean forwards and keep your head to one side. Insert your thumbs into the patient's waistband at the *sides* of the waist

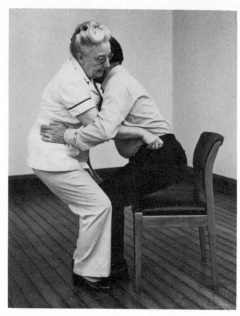

Fig. 85. The waistband lift. Note the ten-to-two stance of the lifter and the forward lean of both people. The lift is in the line of the patient's back.

(see Figs 29 and 85). Keep your arms slightly bent and elbows out. If the patient is reasonably stable and can help, your arms should be inside his or hers. If the patient has little stability and cannot or will not help put your arms outside his or hers and let the hands rest on your waist.

Commands I'M GOING TO PULL ON YOUR WAISTBAND. NOW STAND UP.

Execution Pull on the patient's waistband in a forward and upward direction and in the line of the back, straightening your own legs and slightly bending your elbows. As the patient rises move one hand back to rest over the sacrum and push him or her forward towards yourself then move your other hand back to between the shoulder blades (see Fig. 90). Wait to allow the patient to balance. You may need to take a step back with one of your feet if the patient is

either
Very obese

or
Very tall.

Belt Lift

CHAIR TO STANDING USING A BELT (Figs 86–88)

If the patient has no waistband then a belt can be fastened

either
Around the waist—a plain leather or strong plastic belt—in which case proceed as for the waistband lift

or
Around the hips—a long canvas or other material belt with hand loops (Fig. 88). This is a more advanced lift for a patient who needs only a little help.

Position the Equipment

Fix the chair firmly. Fasten the waist belt firmly with the buckle at the back or fix the canvas belt so that both turns are high under the thighs. The two ends should be at the same level on each side of the patient (Fig. 88).

simultaneously

Inform and Position the Patient

Ask the patient to move to the front of the chair, feet together and hands either on the lifter's waist (see Fig. 85) or forearms resting on the lifter's shoulders—but not if the lifter is very tall (Fig. 86).

Position of the Lifter

Stand in front of the patient, one foot in advance of the other, your hands grasping the belt at each side and your hips and knees bent outwards (Fig. 87)

| **Commands** | I'M GOING TO PULL ON THE BELT. NOW STAND UP. |

| **Execution** | On command the patient stands, aided by a firm, smooth and equal pull on both sides of the band by the lifter. Once the patient is standing, drop the belt letting it fall on the floor and stabilize the patient using either the grasp shown in Figure 91 or the palm-to-palm thumb grasp (Fig. 90). |

| **The Belt for the Belt Lift** | The belt may be made of any firm soft material and should be about 60–70 mm wide and 1,400 mm long. A folded length is more satisfactory and zigzag stitching will help to keep it straight and uncreased. The |

Fig. 86. *(left)* The belt lift, for use when there is no belt on the clothing. The belt should be as high as possible on the thighs. The patient's arms are in an optional position to those in Figure 85.

Fig. 87. *(right)* The belt lift almost completed.

                                    Belt Lift

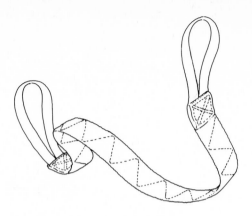

Fig. 88. The belt used in the belt lift.

handles consist of two thinner strips of double material 30 mm wide and 800 mm long inserted into the ends of the wider band. The material should be washable (Fig. 88). Alternatively a firm, non-elastic bandage may be used.

Forearm Support Lift

SITTING TO STANDING (Fig. 89)

The forearm support lift is intended for a patient who needs only a little help to stand up. The patient may possibly be able to get up from a chair with arms but not from an armless chair.

Position the Equipment
Fix the back of the chair against something solid.

simultaneously

Inform and Position the Patient
Ask the patient to move to the front edge of the chair and to separate his or her feet. Tell the patient to grasp your elbows and push down on them as he or she rises.

Position of the Lifter
Stand in front of the patient with your feet in the walk position. Your forward foot may need to be against one of the patient's feet if the floor is slippery. Bend your hips and knees and keep your weight on your back foot. Turn your palms and forearms upwards (supination) and put your forearms under the patient's, grasping under the elbows so that he or she can grasp your elbows (see Fig. 31).

Command
PUSH DOWN ON MY ELBOWS AND STAND UP.

Execution
As the patient pushes *down* on your elbows, you push *up* on the patient's elbows so that you assist the rise into standing. As the patient stands up you straighten your hips and knees and transfer your weight equally on to both your feet (Fig. 89).

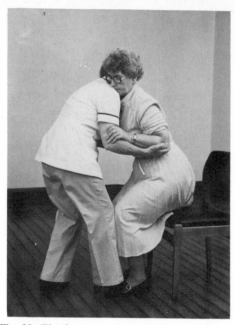

Fig. 89. The forearm support lift. Note the walk stance and the low level of the lifter's arms so that the patient can be pushed up.

Special Points

Do not let go until the patient is balanced. If the patient starts to lose balance

either

Sit him or her down quickly by grasping the elbows and pulling down and back on them

or

Use the double hand pressure shown in Figure 91 by moving one hand first on to the sacrum and then the other to between the shoulder blades and, if necessary, move your feet to the ten-to-two position.

Support in Standing

Support from in Front

(following waistband or belt lift)
The lifter's feet should be in the ten-to-two stance. As the patient stands up move one of your arms first to the sacrum to push the patient forward against your body. Your other arm is then moved to between the shoulder blades and also pushes the patient forward against you (Fig. 91). A patient supported like this is very securely held.

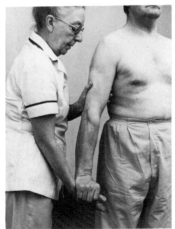

Fig. 90. *(left)* The standing support using the palm-to-palm thumb grasp.

Fig. 91. *(right)* The standing support used following a waistband or belt lift. The lower hand is over the sacrum and is always moved there first and immediately the lift has been completed.

**Support from the
Side**

Using the Palm-to-Palm Thumb Grasp
(following any other lift).

You may have to do the palm-to-palm thumb
grasp by moving from in front of the patient
where you may have been standing to help
him or her to stand up following the forearm
support lift or waistband or belt lift. If this is
the case, first move the foot backwards which
is furthest away from the side on which you
will eventually stand (most probably the
patient's weakest side), then move your other
foot. At the same time, take hold of the
patient's hand in the palm-to-palm thumb
grasp (Fig. 90) using your forward hand, and
use the relaxed hand grasp with your rear
hand to brace the elbow of the same arm by
pushing it forward, that is, into extension.
Tell the patient to push down on to your
hand. A second helper may take up a similar
grasp on the patient's other side or the
patient's walking aid may be used.

Turning in Standing

Stand in front of the patient and hold him or her closely using the grasp shown in Figure 91—one hand on the sacrum and one between the shoulders. Your feet should be in the ten-to-two stance holding the patient's feet between your own feet.

If you wish to turn to your own right, gently rock the patient to your left and at the same time move your right forefoot away from his or her feet. Slightly pivot your body to the right and the patient will turn. Rock the patient upright and back to your left and continue as above.

This manoeuvre can be performed through any number of degrees 90°, 180° or 360° and, indeed, used to be a dance performed in circumscribed spaces. It is suitable for use on heavily handicapped patients who cannot cooperate but whose lower limbs can be straightened. It is advocated for use by family members on a very disabled relative, but is increasingly used for mentally and physically handicapped patients.

To Sit a Patient Down

Position the patient so that the back of the
lower legs both touch the front of the seat.
Use the ten-to-two stance to block the
patient's feet with your feet. Place your hands
palm downwards and thumbs in front of each
side to the patient's waist. Push firmly
downwards on to the pelvis exerting slightly
more pressure with your thumbs. The patient
should easily sit down.

Lifting a Patient from Lying to Sitting Up

(Figs 92–93)
Before a patient sits up it is essential that he or she is told first

either
To take a deep breath in and out

or
To pull the tummy in

or
To raise and lower the head

This causes the abdominal muscles to contract.

Any of these actions will redistribute blood and prevent a feeling of giddiness or faintness.

There are several ways of lifting a patient from lying to sitting up. The patient may help, the lifter may do all or almost all of the work and sometimes two lifters are needed as the patient may do nothing.

A1. One Lifter— Patient Helping— Both Hands

Stand in walk standing at the side of the bed and facing the head of the bed. Offer the patient both your hands to grasp. Keep your hands low near the patient's thighs. Tell the patient LIFT YOUR HEAD AND PULL. Both of you pull and you transfer your weight on to your rear leg as the patient sits up.

A2. One Lifter—
Patient Helping—
One Hand

The lifter stands in walk standing on the opposite side from the patient's able hand. Offer your nearside hand across to the patient's other side and with your outside hand, first put the disabled arm to lie obliquely across the body, then put your hand behind that shoulder using the relaxed hand grasp. Tell the patient LIFT YOUR HEAD AND PULL. Both of you pull, and you use your hand on the shoulder only to steer the patient upright and straight. Again transfer your weight backwards as you pull.

A3. One Lifter—
Patient Helping
but Not Grasping

Stand with feet apart at the side of the bed. Help the patient to rest a hand on each thigh. Lean sideways towards the head of the bed, bending that side knee and hip and put your hand and arm under the neck and across the upper back towards the opposite side of the patient's chest. With your other hand grasp the bed frame (Figs 92 and 93). Tell the patient LIFT YOUR HEAD AND PUSH YOUR HANDS DOWN YOUR THIGHS. At the same time

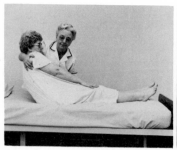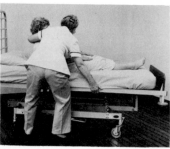

Fig. 92. *(left)* Assisting a patient from lying to sitting up.

Fig. 93. *(right)* The lifter's stance when assisting a patient from lying to sitting up.

Lying to Sitting Up

transfer your body weight from your foot near the head of the bed to that near the bed foot and the patient is assisted to sit up.

A4. One Lifter— Patient Not Helping

Follow the procedure in A3, but cradle the patient's head on your shoulder as you insert your arm behind the neck and upper back. As the patient is sat up ensure that you keep the head and upper trunk supported against your body. This move can be performed unexpectedly fast, so do be careful.

B1. Two Lifters—Patient Helping

Both lifters stand in walk standing facing the bed head. They put their inside hands under the patient's axilla to rest on the back of the shoulders. The patient raises both arms and grasps the lifters' elbows or shoulders in the same way. Tell the patient LIFT YOUR HEAD AND PULL. All three pull, the lifters transfer their weight on to their back legs and use their outside hands to steady the patient.

B2. Two Lifters—Patient Not Helping

The method described in B1 may be used but both lifters also put their outside hands under the patient's head which is raised first. Then proceed as in B1.

or

Method A3 is used, but both lifters each insert an arm under the head and upper thorax. Only a very heavy and very helpless patient needs this method. Also, like A4, this move can be unexpectedly fast, so do be careful.

Lying to Sitting Up

Lifting and Moving Loads Other Than Patients

When lifting and moving nonhuman loads, many of the principles that have already been outlined apply, but the variations are fewer. Before lifting or moving remember first the following points on feet, posture, grasp and arms.

Feet   Your foot positions should give you a good stance near to the load and a wide enough base with one foot pointing in the direction of the move.

Posture   Use the same rules of keeping your back erect and head up, with the movement occurring at your hips, knees and ankles.

Grasp   Use as much of your hand as you can to support an object and never less than all your fingers. Always try to grasp under the load with one of your palms upturned. The other hand may also go under the load or be at the side to stabilize.

Arms   Keep your arms close to your body with your elbows tucked well in.

ADDITIONAL PREPARATION FOR EVERY MANOEUVRE

Practise Bracing   Do a dynamic abdominal brace and take up your own body slack by raising your head and stretching yourself a little.

Pull or Push	Pulling is less stressful than pushing and you should always try to pull beds, stretcher trolleys, gas cylinders, medication and equipment trolleys. If they have handles use them. Turn the casters so that they wheel instead of resisting. Leaning forwards to push increases your intra-abdominal pressure more than leaning slightly backwards to pull.
High Loads	When lifting from high shelves, get a stool or step ladder rather than overreach. Getting heavy loads of linen or heavy or awkward equipment off high shelves is stressful both during the high reach and during the lowering action so try to get your arms level with the shelf.
Low Loads to High Shelves	Arrange an intermediate platform to be used as a staging post. First move the load from the floor to this point. Re-adjust your position, grasp and continue the lift.
Tucking in Bedclothes	Everytime bedclothes are tucked in you can increase your intra-abdominal pressure, and this is certainly a difficult manoeuvre if you suffer from back pain. Try to stand in the walk or the lunge position. Always brace your abdomen, lift the mattress with one hand and, at the same time, lift your head back and up *before* tucking the bedclothes with the other hand. If you are making a low bed, kneel down beside it if the bed cannot be elevated.
Heavy Loads	Don't be ambitious—split the load, make two trips or get help.

Bibliography

BARTELINK D. L. (1957) The role of abdominal pressure in relieving the pressure on the lumbar intervertebral discs. *J. Bone Joint Surg.* **39B,** 718–25.

BELL F., DELGITY M. E., FENNELL M. J. & AITKEN R. C. B. (1979) Hospital ward patient lifting tasks. *Ergonomics* **22, 11,** 1257–73.

CUST G. (1972) The prevalence of L.B.P. in nurses. *Int. Nursing Review* **19, 2,** 169–72.

DAVIS P. R. (1956) Variations in intra-abdominal pressure during weight lifting in various postures. *J. Anat.* **90,** 601(p).

—— (1965) The nurses' load. Leading article, *The Lancet* **ii,** 422–23.

—— & STUBBS D. A. (1978) *A method of establishing safe handling forces in working situations.* DHEW (NIOSH) Pub. No. 78–185, 34–8.

—— STUBBS D. A. & RIDD J. E. (1977) Radio pills: their use in monitoring back stress. *J. Med. Eng. & Tech.* **1, 4,** 209–12.

—— & TROUP J. D. G. (1964) Pressures in the trunk cavities when pushing, pulling and lifting. *Ergonomics* **7, 4,** 465–74.

DHSS (1977) *Working group on back pain.* London: HMSO.

HOLLIS M. (1979) Prevention means practising. *Proc. Conf. on Prevention of Back Pain in Nursing,* Northwick Park Hospital and Nursing Practice Research Unit, 28–30.

—— & WADDINGTON P. J. (1975) *Handling the Handicapped.* Cambridge: Woodhead-Faulkner.

KENNEDY B (1980) An Australian pro-
gramme for management of back problems. *Physiotherapy* **66,** 4, 108–11.

PHEASANT S. T. (1979) The biomechanics of
the human spine. *Proc. Conf. on Prevention of Back Pain in Nursing,* Northwick
Park Hospital and Nursing Practice
Research Unit, 9–13.

ROYAL COLLEGE OF NURSING (1979)
Avoiding Low Back Injury Among Nurses.
London: Rcn.

STUBBS D. A., HUDSON M. P., RIVERS P. M.
& WORRINGHAM C. J. (1979) Patient
handling and truncal stresses in nursing.
*Proc. Conf. on Prevention of Back Pain in
Nursing,* Northwick Park Hospital and
Nursing Practice Research Unit, 14–27.

STUBBS D. A. & OSBORNE C. M. (1979) How
to save your back. *Nursing* **3,** 116–24.

TROUP J. D. G. (1979) Causes of back pain at
work and the mechanics of back injury.
*Proc. Conf. on Prevention of Back Pain in
Nursing,* Northwick Park Hospital and
Nursing Practice Research Unit, 2–7.

Index